GEOMETRY
FOR THE PRACTICAL WORKER

MATHEMATICS LIBRARY FOR PRACTICAL WORKERS

Arithmetic *for the Practical Worker*
Algebra *for the Practical Worker*
Geometry *for the Practical Worker*
Trigonometry *for the Practical Worker*
Calculus *for the Practical Worker*

A Group of Books That Make Easy
the Home Study of the Working
Principles of Mathematics

GEOMETRY
FOR THE PRACTICAL WORKER

4th Edition

J. E. Thompson

MATHEMATICS LIBRARY FOR PRACTICAL WORKERS

VAN NOSTRAND REINHOLD COMPANY
NEW YORK CINCINNATI TORONTO LONDON MELBOURNE

Menominee County Library
Stephenson, Michigan

Copyright © 1982 by Van Nostrand Reinhold Company Inc.

Library of Congress Catalog Card Number: 81-11690
ISBN: 0-442-28272-9

All rights reserved. Certain portions of this work copyright © 1962, 1946, 1931 by Van Nostrand Reinhold Company under the title, *Geometry for the Practical Man*. No part of this work covered by the copyright hereon may be reproduced or used in any form or by any means—graphic, electronic, or mechanical including photocopying, recording, taping, or information storage and retrieval systems—without permission of the publisher.

Manufactured in the United States of America

Published by Van Nostrand Reinhold Company Inc.
135 West 50th Street, New York, N.Y. 10020

Van Nostrand Reinhold Limited
1410 Birchmount Road
Scarborough, Ontario MIP 2E7, Canada

Van Nostrand Reinhold Austrailia Pty. Ltd.
17 Queen Street
Mitcham, Victoria 3132, Australia

Van Nostrand Reinhold Company Limited
Molly Millars Lane
Wokingham, Berkshire, England

15 14 13 12 11 10 9 8 7 6 5 4 2

Library of Congress Cataloging in Publication Data

Thompson, James Edgar, 1892-
   Geometry for the practical worker.

   (Mathematics library for practical workers)
   Includes index.
   1. Geometry. I. Peters, Max, 1906-
II. Title. III. Series.
QA453.T5   1981   516   81-11690
ISBN 0-442-28272-9   AACR2

PREFACE TO FOURTH EDITION

The popular demand for this book and for the other four in the MATHEMATICS LIBRARY FOR PRACTICAL WORKERS has been maintained through three editions and for over fifty years. The publisher hopes that this Fourth Edition will prove as valuable to a new generation of practical workers who must work with mathematics on a regular, practical, nontheoretical basis.

PREFACE TO THIRD EDITION

IN this Third Edition of *Geometry for the Practical Man*, as in the revisions of the other volumes in this series, the authors have made every effort to retain the basic methods of presentation which have proven so successful in the earlier editions. However, new concepts have been introduced in the subject matter which, it is believed, will better prepare the student to study and to understand those parts of advanced mathematics in which he may be interested or in which he may feel the need to acquire competence. The non-Euclidean geometries play a major part in the intellectual and mathematical history of our time, and a chapter on them has been added to introduce the student to at least one of their basic postulates as well as to give him a modest appreciation of their significance. Moreover, in order that the student may further test his proficiency, a new section of Review Exercises has been added.

PREFACE TO FIRST AND SECOND EDITIONS

SIR ISAAC NEWTON, the English creator of the calculus and discoverer of the law of gravitation, studied geometry without the aid of a teacher and alone, but for most students it has long been considered a difficult subject and its teaching has been considered as belonging exclusively in the schools and organized classes. In recent years, however, there has been let loose on the general public a flood of books which attempt to tell the general reader "all about science," and a few such books on mathematics have timidly felt their way about in the flood. Most of these have been more about mathematics than of mathematics, and in particular they have not often tackled the ancient science of geometry. Furthermore, most books on so-called popular science and mathematics are either superficial and do not really teach, or they omit much of a subject because they use their available space in trying to delve into a few parts of the subject.

The first edition of this book, written as one of a series of similar treatments of other branches of mathematics, attempted to give a balanced treatment of the usual parts of elementary geometry in such form that it could be studied by any reader having a knowledge of arithmetic and a little elementary algebra, without the aid of a teacher. The very large sale which the publishers found for the book indicated either that its plan and method were successful or that a surprisingly large number of people were waiting for a book which they could use without a teacher. The publishers and the author have received a very large correspondence from readers which shows a great interest in and satisfaction with the book and indicates some small changes which seem necessary. In order to conform more nearly to the original plan some of the more formal proofs and calculations have been omitted in this edition and replaced by descriptive explanations and illustrations.

As in the other books of the series one of the aims here is to present the fundamentals of elementary geometry from the viewpoint of those who would wish to use it for practical purposes in the arts and industry and in the study of other branches of mathematics. Thus, while not a book on so-called "practical geometry" as such, it is intended that the book shall be useful to those who wish to *apply* the *principles* in any way.

In conformity with the plan as stated the treatment has been made as informal as the nature of the subject will allow. Some of the traditional parts of the subject have been omitted, and several things have been included which are not usually found in books on elementary geometry. In order to indicate to the reader that the subject of geometry in its modern form is the result of a long and very human effort, rather than a gift from the gods in completed and perfect form, a fairly comprehensive sketch of its history is given in the first chapter. Because of the antiquity of the science of geometry, its fundamental importance in the study of other branches of mathematics, and its very great technical and practical importance, and also because it is the bright and shining example of the methods of logical procedure, the historical sketch is made long enough to be of some real value to the interested reader. References to this chapter and supplementary historical notes on important individual topics are given at appropriate places in the text in order to supply additional information or to make clear allusions found in mathematical literature.

The introduction to the main facts of geometry has been put into intuitive form and each new subject is introduced in the same manner or by means of its connection with one which has been previously discussed. A large number of illustrations and applications have been given; these are really more illustrations than applications, however. In order to understand and appreciate *applications* of mathematics one must have already some technical knowledge of the subject in which the application is made and such knowledge has not been assumed here, although this may be done in the higher branches of mathematics. A short *Foreword* has been included for readers who here approach the subject for the first time. Although the long and severely formal proofs have been omitted the simpler proofs are given fairly complete, though not in formal style, in order to illustrate and cultivate systematic reasoning and logical processes. The numerical parts of the subject, amounting to applications of some of the principles, are an extension and amplification of the work of arithmetic on measurement.

J. E. THOMPSON

Brooklyn, N. Y.
October, 1945

CONTENTS

PART I

INTRODUCTION TO GEOMETRY

CHAPTER 1 ORIGIN AND DEVELOPMENT OF GEOMETRY

ART.		PAGE
1.	BEGINNINGS OF GEOMETRY—THE BABYLONIANS	3
2.	THE ANCIENT EGYPTIANS	3
3.	GEOMETRY AND THE EARLY GREEKS	4
4.	EUCLID AND ARCHIMEDES	7
5.	THE LATER GREEKS IN EGYPT	9
6.	GEOMETRY AMONG THE ARABS AND HINDUS	11
7.	GEOMETRY AND THE UNIVERSITIES IN THE MIDDLE AGES AND SINCE	12
8.	MODERN DEVELOPMENT OF GEOMETRY	14
9.	NON-EUCLIDEAN GEOMETRY	15

CHAPTER 2 SOME PRELIMINARY IDEAS AND METHODS OF GEOMETRY

10.	WHAT IS GEOMETRY?	16
11.	DIVISIONS OF GEOMETRY	17
12.	METHODS OF EUCLIDEAN GEOMETRY	18
13.	STRAIGHT LINE AND RULER	18
14.	CIRCLE AND COMPASS	20
15.	SOME FUNDAMENTAL DEFINITIONS	22
16.	GENERAL TERMS USED IN GEOMETRY	23
17.	THEOREMS AND THEIR PROOF	24
18.	PROBLEMS AND THEIR SOLUTIONS	25
19.	STARTING POINTS OF GEOMETRY	26
20.	GENERAL POSTULATES	27
21.	THE GEOMETRICAL POSTULATES	28

PART II

PROPERTIES OF PLANE FIGURES

CHAPTER 3 SOME PROPERTIES OF STRAIGHT LINES AND ANGLES

22.	STRAIGHT LINES	31
23.	ANGLES	32

CONTENTS

ART.		PAGE
24.	Perpendicular Lines and Right Angles	34
25.	Other Angles and Their Relations	35
26.	Extension of the Meaning of Angles	36
27.	Measurement of Angles	38
28.	Exercises	40
29.	Some Properties of Perpendicular and Oblique Lines	40
30.	Drawing of Perpendiculars	44
31.	Parallel Lines	46
32.	Some Further Properties of Angles	50
33.	Drawing of Angles and Parallels	51
34.	Size of the Earth	54
35.	Exercises	56

Chapter 4 PLANE FIGURES FORMED BY STRAIGHT LINES AND ANGLES

36.	Introduction	58

A. Triangles

37.	Forms of Triangles	59
38.	Triangle Descriptions and Definitions	61
39.	Some General Properties of Triangles	63
40.	Some Properties of Altitudes, Medians, etc., of Triangles	68
41.	Right Triangles	70
42.	Triangle Constructions	73
43.	Illustrations and Applications	76
44.	Exercises	84

B. Quadrilaterals

45.	Forms and Descriptions of Quadrilaterals	85
46.	Some Properties of Quadrilaterals	87
47.	Quadrilateral Constructions and Applications	94
48.	Exercises	96

C. Polygons

49.	Definitions and Descriptions	97
50.	Properties of Polygons	99
51.	Exercise	103

Chapter 5 SOME PROPERTIES OF THE CIRCLE

52.	The Circle	104
53.	Definitions Pertaining to the Circle	105
54.	Properties of Arcs, Chords, and Tangents	107
55.	Measurement and Limits	113
56.	Measure of Angles	115
57.	Circle Constructions	120
58.	Illustrations and Applications	125
59.	Exercises and Problems	131

CONTENTS xi

CHAPTER 6 PROPORTION AND SIMILAR FIGURES

ART. PAGE
60. RATIO AND PROPORTION 133
61. PRINCIPLES OF PROPORTION 134
62. PROPORTIONAL LINE SEGMENTS 135
63. SIMILAR FIGURES 137
64. PROPERTIES OF SIMILAR TRIANGLES 138
65. CIRCLE PROPORTIONS 142
66. CONSTRUCTION PROBLEMS 144
67. ILLUSTRATIONS AND APPLICATIONS 146
68. EXERCISES AND PROBLEMS 155

PART III

MEASUREMENT OF GEOMETRICAL FIGURES

CHAPTER 7 DIMENSIONS AND AREAS OF PLANE FIGURES

69. INTRODUCTION 161
70. DIMENSIONS OF PLANE FIGURES 161
71. AREA AND ITS MEASURE 162
72. AREAS OF RECTANGLES AND PARALLELOGRAMS ... 163
73. REMARKS ON AREA MEASURE 167
74. AREAS OF TRIANGLES 167
75. NOTE ON THE PYTHAGOREAN THEOREM 173
76. AREAS OF TRAPEZOIDS AND POLYGONS 174
77. ILLUSTRATIVE EXAMPLES 177
78. EXERCISES AND PROBLEMS 180

CHAPTER 8 REGULAR POLYGONS AND
MEASUREMENT OF THE CIRCLE

79. REGULAR POLYGONS AND CIRCLES 182
80. RELATIONS OF REGULAR POLYGONS AND CIRCLES .. 183
81. INSCRIPTION OF REGULAR POLYGONS 189
82. LENGTHS OF SIDES OF INSCRIBED REGULAR POLYGONS . 191
83. REMARKS ON INSCRIPTION OF REGULAR POLYGONS AND THE DIVISION
    OF THE CIRCLE 192
84. MEASUREMENT OF THE CIRCLE 192
85. NOTE ON THE NUMBER π 195
86. ILLUSTRATIVE EXAMPLES 197
87. EXERCISES AND PROBLEMS 200

CHAPTER 9 SURFACE AREAS AND VOLUMES
OF SOLIDS HAVING PLANE SURFACES

88. LINES AND PLANES IN SPACE 203
89. SOLIDS 207
90. SOLIDS WITH PLANE SURFACES 208

CONTENTS

ART.		PAGE
91.	Rectangular Solids and Volume	208
92.	Prisms and Pyramids	213
93.	Regular Solids	216
94.	Note on the Regular Solids	219
95.	Exercises and Problems	221

Chapter 10 THE "THREE ROUND BODIES" AND THEIR MEASUREMENT

96.	Solids with Curved Surfaces	223
97.	The "Three Round Bodies"	224
98.	The Cylinder	225
99.	The Cone	228
100.	The Sphere	230
101.	Exercises and Problems	232

Chapter 11 NON-EUCLIDEAN GEOMETRY

102.	Non-Euclidean Geometry	234
103.	The Parallel Postulates	235
104.	The Sum of the Angles of a Triangle	238
105.	Visual Representation of Non-Euclidean Geometry	243

Review Exercises in Geometry	245
Answers to Problems	252
Answers to Exercises	255
Index	259

FOREWORD

THE question is often asked: Why should one study geometry? The answer depends on the questioner.

For the practical man it may be said that geometry has very wide application in the arts and trades, such as building and architecture, machine work, surveying, engineering, etc. This does not necessarily mean that the worker consciously applies the methods of geometry, rule by rule and principle by principle, but that the rules and methods which he does use in his work are based on or derived from the principles and methods of geometry.

For the student who intends to go farther in the study of mathematics and science and enter the higher branches it is one of the necessary foundations, and of an importance which requires no urging.

In this connection it may be said that after we learn to count and perform the simple operations of arithmetic, algebra and geometry form the joint foundations of all pure and applied mathematics. Algebra, as remarked elsewhere, supplies the "alphabet and grammar," so to speak, and geometry develops the forms and methods of reasoning used in mathematics. The one lays the foundations of the rules and methods of calculation, and the other the foundation of the rules and methods of measurement; and measurement and calculation are the heart of pure and applied mathematics.

When all is said and done, however, the fact remains that geometry has attracted and held the attention and interest of the human race from prehistoric times until the present, purely for its own sake. Geometry is at the same time a science and an art, mathematics and philosophy. It supplies us the only perfect system of logic and its beautiful completeness is not to be found in any other branch of knowledge. For one who loves logic, completeness, perfection, and beauty, geometry is a subject which is always fascinating, and which supplies its own reason for existence.

In this book we cannot set forth geometry in all its completeness and perfection, and in the procedure which we must follow it will unavoidably seem to lose some of its beauty. To one seeking to find something of its meaning and its usefulness, however, it can never lose its interest.

The method which is followed in this book is stated in the preface as being threefold: to make the study of geometry possible without a teacher, to present those parts of the subject which will be most directly useful in its applications and in further study, and to reveal something of its eternal interest. As an aid toward this last object a chapter on the history of geometry is given at the beginning of the book. It is hoped that this chapter will serve to open up to the reader approaching the study for the first time, the long road and vista down which geometry has come to us, beginning and progressing with the history of our race.

In the study of geometry many of the rules, methods and symbols of arithmetic and algebra are used and it will be assumed in this book that the reader has some knowledge of these subjects. In other wordy, in the usual course of study in mathematics, geometry comes after arithmetic and algebra.*

When studied as a compulsory subject and in the manner in which it is sometimes approached and presented geometry can be made distressingly dry and dull, but to one who begins its study voluntarily, and when it is properly approached and handled, geometry will be found to be simple and fascinating, and its mastery will provide the highest intellectual satisfaction and great practical utility.

In the statement of geometrical facts and in the proofs of new geometrical relations from those already established it is necessary and customary to refer to those previous relations, and this is done by giving each important result a number. This number is then cited as reference whenever it is used. In this way it comes about that any book on geometry will apparently require on the part of the reader frequent turning back to earlier results. This is not necessary in ordinary reading in this text, however, as most of these previous results are mentioned by quoting briefly their most important features. Thus the otherwise forbidding appearance of so many references need cause no trouble and these may be ignored except when specially needed or desired. In such cases the number of the statement and the article

---

* *Note.*—All the knowledge of algebra necessary for the study of this book may be obtained from the author's "Algebra for the Practical Man," published by D. Van Nostrand Company, Princeton, N. J.

in which it is found are available. In general, while some of the formality of strict logic is unavoidable, and occasional use of algebra is necessary, it is hoped that the reader will find this book easy and informal reading and at the same time a sufficiently complete treatment of the ancient and perennially fascinating subject of GEOMETRY.

Part I
INTRODUCTION TO GEOMETRY

Chapter 1

ORIGIN AND DEVELOPMENT OF GEOMETRY

1. Beginnings of Geometry—The Babylonians. Prehistoric man probably had only the vaguest notions of number and size and very likely did his first counting on his fingers and his first measurement by comparing lengths with the lengths of certain parts of his body, such as the foot, the arm, or the outstretched arms. In caves in certain parts of Europe, protected from the weather, have been discovered drawings made in prehistoric times which reveal a fairly precise sense of dimensions and proportion but the artists and draftsmen left no records of their methods or systems of measurement.

The earliest indications of any system of measurement seem to come from the ancient Babylonians or the people who lived before them in the region known as Babylonia. The Babylonians developed a system of land measurement and their clay tablet records show that they had methods for finding the areas of several simple figures, including the circle, though their ideas of circle measurement were not quite correct.

Most of the ancient records show that definite methods and knowledge of measurement arose in connection with land measurement, building, and astrology, the pseudo-science of prediction by study of the heavenly bodies, which was the forerunner of astronomy. In this connection the Babylonians supposed that the heavens revolved around the earth and that the year consisted of 360 days. This led them to divide the circle into 360 parts and thus probably originated the present *degree* system of angle measure.

The measurement of simple figures formed of straight lines and circles required a knowledge of their properties, and it was the study of these properties which led to the development of the science of geometry.

2. The Ancient Egyptians. In connection with the building of the pyramids, the surveys of the valley of the Nile following the annual floods, and the pseudo-science of astrology, the ancient Egyptians built up a considerable body of knowledge of arithmetic, simple algebra and measurement. The records of this knowledge have been preserved and brought down to us in the ancient manuscripts discovered in modern times in the pyramids, tombs and palaces of the early Egyptians.

The oldest of these ancient Egyptian manuscripts which have so far been discovered was written by one Ahmes, a priest, about 1700 B.C. The original manuscript is now in the British Museum. It consists of a sort of summary or collection of rules and problems with their answers, dealing with arithmetic and the measurement of various geometrical figures. The title of the treatise or summary is, "Directions for Knowing all Dark Things," which indicates that mathematical knowledge was a secret which belonged to the priesthood and those intrusted with the erection of public works. This work seems to be a copy, with improvements, of an earlier work which dates back about a thousand years. Thus the earliest written record of definite geometrical knowledge dates back to 1700 B.C. and probably to about 2700 B.C.

The fame of the wisdom of the Egyptians spread over all the civilized world of their time and students and scholars came from other countries to travel and study in Egypt. Among these were the ancient Greeks, who began to come to Egypt about 600 B.C. and were much impressed by the Egyptian system of land measurement and calculation. They studied this system carefully and gave to it the simple name "gemetrein" (Greek, $\gamma\hat{\eta}$ (ge) = "earth"; $\mu\epsilon\tau\rho\epsilon\hat{\iota}\nu$ (metrein) = "to measure"), which has come down to us as *geometry*.

3. Geometry and the Early Greeks. With the ancient Egyptians their system of land measure was purely practical and of interest only so far as it was useful, and they paid but little attention to its development beyond the point of usefulness in the practical arts. The early Greek philosophers, on the other hand, were interested in the knowledge of geometry purely for its own sake, regardless of utility, and they studied, developed and extended it as a branch of learning, purely through intellectual curiosity and love of knowledge.

The Greek scholars returned from Egypt to their own country and taught geometry in their private schools, and a great interest in the new knowledge grew and spread among the Greek philosophers. They studied the properties of the geometrical figures, the relations among these properties, and the proof by pure logic of new geometrical truths from those already known. In this way they soon knew far more about the subject than the Egyptians and wrote papers and books dealing with various parts of the subject.

The Greek surveyors, builders, mariners and astronomers soon began to make use of the practical parts of the philosophers' pure science of geometry and applied it to surveying, architecture, sculpture, navigation and astronomy. Those things which the Egyptians had done by means of geometry the Greeks now did much better, and they also extended the use of geometry to other fields where the Egyptians had not made use of it. These applications of geometry the Greeks developed into fine arts and sciences, and from these sciences and geometry together came the new sciences of trigonometry ("trigon measure," *trigon* being a *triangle*) and of geodesy, which has to do with the size and shape of the earth as a whole.

The Greek scientists knew that the earth is a very nearly round ball and determined very closely its size and exact shape. One of the leaders in this work was Eratosthenes, who in later years went to Egypt to teach, after that country was conquered by Alexander the Great. He determined the circumference and diameter of the earth to within a few hundred miles, determined almost exactly the length of the year (time of revolution of the earth), and suggested the calendar now known as the *Julian Calendar* (see the ARITHMETIC of this series), which was used until comparatively recent times.

The Romans, who finally conquered the Greeks in a military sense, took over the practical geometry used in building, navigation and engineering, but paid little attention to pure geometry as a science and contributed nothing to its development.

Of those ancient Greeks who brought back from Egypt a knowledge of geometry, the earliest of whom any individual record survives was Thales, who was born in the town of Miletus in 640 B.C. and died in the same place at the age of ninety. He was a wealthy business man of a very practical turn of mind and apparently first visited Egypt in middle age on a business trip. While there he became interested in geometry as a practical art. He remained in Egypt for some time and soon after his return to his native city he retired from business and devoted the remainder of his life to the study and teaching of geometry and astronomy. It is as a geometer rather than as a business man that he is now famous. One of the pupils of Thales, Anaximander, is said to be the first who attempted to classify the various parts of the subject of geometry and to write a book on the subject.

The first to investigate systematically the principles upon which the science of geometry is based and to apply the methods of logic to its systematic development was Pythagoras (569–500 B.C.), who first studied under Anaximander and then went to Egypt. After some years in Egypt he settled in a Greek colony in the south of Italy and taught geometry, philosophy and religion, attempting to base the last two subjects on mathematical principles. His school grew into a sort of brotherhood which finally developed into a secret society or fraternity. The emblem of the society was the five-pointed star, drawn by continued motion of the pen without lifting it from the paper, as shown in the accompanying figure. At the corners of this figure were placed the letters of the Greek word υγιθα (ugitha) meaning "health." In the study of the properties of this figure the Pythagoreans discovered also many of the properties of triangles (three-sided figures) and pentagons (five-sided figures).

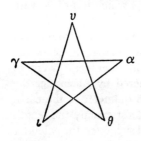

Pythagoras taught other branches of mathematics beside geometry and in geometry discovered several important new rules and propositions beside those already known. In their systematic development of the subject he and his followers gave either the first or better proofs of a number of previously known propositions. Of these he is said to be the first to prove that the square constructed with a side equal to the longest side of a triangle with one right angle (square corner) has an area equal to the sum of the areas of the squares whose sides are the other two (shorter) sides of the triangle. In his honor this proposition is known as the *Pythagorean Theorem.* Many other proofs of this theorem besides that of Pythagoras are now known.

Another of the famous early Greek Geometers was Archytas (born about 400 B.C.) who was a follower of the Pythagoreans. He was the first to solve the famous problem of determining by geometry the dimensions of a cube which should have twice the volume of any specified cube, called the "duplication of the cube." Archytas is said also to be the first to make use of the principles of geometry in the study of mechanics.

Hippias of Elis, who lived at about the same time as Archytas, was the first to solve the theoretical problems of dividing an angle into three equal parts ("trisection of an angle") and of "squaring the circle," or finding a square which shall have the same area as any specified circle. The solution of these famous problems was not accomplished by the methods of the simple pure geometry however (and is indeed impossible by such methods), and among the leaders in the search for such methods of solution was Hippocrates of Chios, who was born about 470 B.C. (not the celebrated physician Hippocrates of Cos, of the same period). In his search for the solutions of these problems Hippocrates discovered many new properties of the circle and of angles, as well as other geometrical truths, and thus extended considerably the knowledge of geometry. He wrote a book on the subject which was the first textbook on pure geometry.

Two of the greatest of the ancient Greek philosophers were Plato (429–328 B.C.) and his pupil Aristotle (384–322 B.C.), both of whom lived and taught in Athens, the capital and center of Greek culture. Plato specialized in pure philosophy and geometry and paid particular attention to the logical foundations of geometry. He conducted a school which was located in a grove or park near the city of Athens and is said to have placed over the entrance a notice which warned away those who knew no geometry. Aristotle took all knowledge as his field and among other things wrote on geometry and physics, coming to be the recognized authority of his time and for centuries afterward.

After Aristotle's time the center of geometrical and other Greek learning shifted back to Egypt, which had by then been conquered and colonized by the Greeks under Alexander the Great. He built on the delta of the Nile a great city, which he named Alexandria for himself, and there founded a great university and library. To the University of Alexandria flocked the scholars and students of all the world, and for hundreds of years it was the center and headquarters of the study and development of geometry.

4. Euclid and Archimedes. Two of the greatest mathematicians of the Alexandrian golden age and, for that matter, of all antiquity, were Euclid, an Alexandrian of Greek parentage, and Archimedes, a Sicilian Greek who studied and worked for some years in Alexandria.

Of Euclid very little is known beyond the facts that he was born about 330 B.C. and died about 275 B.C., that he spent most of his life in Alexandria, and that he taught mathematics there at the University and to private students for many years. It is to him that the proverb "there is no royal road to knowledge" is attributed. It is said that one of his private students was the young prince Ptolemy, son of King Ptolemy of Egypt, and that the prince was not deeply interested in geometry. He asked his teacher if there was no short and easy way to learn geometry and Euclid replied, "Oh, Prince, there is no royal road to geometry." Although little is known of Euclid's life his work fills a large part of the history of mathematics. Many of his students became famous and these have left accounts of his teaching and of his discoveries and writings, and most of his writings have come down to us. He wrote books on a number of scientific subjects but the most famous of his works are those on arithmetic, algebra and geometry, and of these his fame is based chiefly on his books on geometry.

Euclid's great work was called in Greek στοιχῖα and in Latin *elementa* and so is known as the *elements*, the "Elements of Mathematics." It covered parts of arithmetic, theory of numbers, algebra and proportion, and all that was then known of geometry, and each separate part of the subject was written on a different roll or book, all of which were consecutively numbered. The complete work consisted of thirteen books or sections and of these seven were devoted to geometry, Books I, II, III, IV, VI, XI, XII. Book XIII is a sort of supplement which contains some geometrical material. The books on geometry are usually separated from the others and together are known as Euclid's "Elements of Geometry" and often referred to as *Euclid's Elements*.

The *Elements* was written in Euclid's later years and was the first complete book on the subject. It thoroughly systematized and restated the entire subject, carefully stated the foundations of the science, simplified many of the statements and proofs of the propositions, and classified, rearranged and numbered all the foundation principles, definitions and propositions. It also contained much original material due to Euclid himself. It was at once adopted as a standard textbook and later spread to all the world. It has been translated into many languages and has come down to us as Euclid left it, and to this day it is still used as a textbook and also as the model and basis of all other books on the so-called elementary geometry.

In the *Elements* Euclid supposes that the reader can use the ruler and compass and no other instruments are allowed. Thus all the drawings, proofs and solutions are based on and carried out by means of the straight line and circle alone. The geometry thus developed is called *elementary geometry* (from the *Elements*) and also *Euclidean* geom-

ART. 5   ORIGIN AND DEVELOPMENT OF GEOMETRY   9

etry. During and since Euclid's time other methods have been developed but these belong to what is called *higher* geometry. Some of these will be mentioned later in this sketch.

Euclid's *Elements* deals with the so-called *plane* geometry, which treats of figures which can be drawn on a plane or "flat" *surface*, and with those solid figures, or *solids*, which have flat faces or surfaces. The familiar solids, sphere, cylinder and cone (called the "three round bodies"), are not treated by Euclid. This was done by Archimedes.

Archimedes was born in the Greek colonial city of Syracuse in Sicily in 287 B.C. and was killed there in 212 B.C. He seems to have been related to the family of the king or governor of Syracuse and had the best educational advantages which those times offered, being sent to the University of Alexandria where he studied the arts and sciences, particularly mathematics. He was a scientist and inventor and in addition contributed more original material to geometrical knowledge than any other one person. He perfected the measurement of the circle, sphere, cylinder, and cone, and discovered many of their properties and the relations between them. He systematized all this knowledge after the manner of Euclid and developed it by means of the straight line and circle. His books on these subjects have come down to us and they have been added to elementary geometry.

Archimedes made discoveries, wrote and taught, in many other subjects beside geometry, including other branches of mathematics, mechanics and physics. His mechanical ingenuity was astonishing and a great number of most useful inventions and discoveries, many of which are still in use, were made by him. When Syracuse was attacked by the Romans his aid was sought by the authorities and he designed a machine for hurling heavy projectiles which was almost as effective as a cannon. By the aid of his machines the Romans were held at bay for nearly three years. The city was finally starved into submission by being blockaded, and when it was taken Archimedes was killed by a soldier who did not know who he was.

5. The Later Greeks in Egypt. After the time of Euclid and Archimedes the University of Alexandria flourished for about nine hundred years. The great library became the depository of all the learned works and literature of antiquity and the fame of the mathematicians of all nations who gathered there spread to all the world. Among the mathematicians the Greeks led all the rest. Among these was one who ranked with Euclid and Archimedes and was a contemporary of theirs, Appolonius of Perga (260–200 B.C.). Appolonius studied and developed a branch of the higher geometry dealing with the curves and figures known as the *conic sections*, and his work and methods became the basis for modern *analytical* geometry (article 8).

Another of the Alexandrians was Eratosthenes, already mentioned as the mathematician and astronomer who determined the size and shape of the earth. Another is the geometer who is generally credited with the invention of the branch of mathematics called *trigonometry*, an advanced branch of applied geometry. This was Hipparchus of Nicaea. The mathematical science of land surveying, a great advance over the simple land-measure of the ancient Egyptians and based on pure geometry and trigonometry, was perfected by Hero, who was also a noted engineer and the inventor of the first steam engine (about 120 B.C.).

Many mathematicians followed these at Alexandria but the next great geometer flourished in the second century after Christ. This was Ptolemy of Alexandria, a descendant of one of the kings of Egypt; the date of his birth is not known but he died an old man in 168 A.D. Ptolemy wrote books which treated of orthographic and stereographic projection, the foundations of modern *descriptive* geometry (article 8). His greatest work, however, was the application of geometry (and trigonometry) to astronomy. He wrote a great work on mathematics and astronomy in thirteen books (sections) which did for the science of astronomy what Euclid's *Elements* had done for pure geometry. This work became the standard text on astronomy and applied geometry and remained in use for about a thousand years, until the time of Copernicus in the 16th century. The history of the title of this work is interesting and curious. It was originally in Greek and entitled μεγίστα μαθηματικη σύνταξις (*megista mathematika syntaxis*, or "great mathematical syntax"), was translated into Latin as "Megale Syntaxis Megistos" (*the great(est) collection*), was translated by the Arabs as "Al Midschisti" (*the greatest*) and has come down to us as the *Almagest*. In this book, among many other remarkable things, is the first recorded definite use of *degrees*, *minutes* and *seconds* in the measurement of angles, and the determination of the approximate value of the ratio of the circumference to the diameter of the circle.

A hundred and fifty years after Ptolemy flourished Pappus, who further developed the higher geometry and conic sections and discov-

ART. 6 ORIGIN AND DEVELOPMENT OF GEOMETRY

ered propositions which lie close to the foundations of the calculus. By his teaching he also temporarily revived interest in geometry, which was then declining. After Pappus no great original geometer rose among the other famous mathematicians of Alexandria, but one of the native Greeks, Proclus (412–485 A.D.) wrote a sort of history of geometry which contains a commentary on Euclid's *Elements* together with interesting personal comment on the great Alexandrian geometers.

After Proclus the history of the University and the Alexandrians is no longer interesting from the viewpoint of geometry. It came to an end with the destruction of the University and the Library by the Arabs who conquered and captured Alexandria in 641 A.D. It is said that they used the books (rolls) of the Library as fuel for the furnaces in the public baths of the city and that the supply lasted for nearly a year.

6. Geometry Among the Arabs and Hindus. After the destruction of the University and Library at Alexandria the scholars left Egypt and scattered over the entire civilized world. Many of them went to Byzantium (later called Constantinople, after the Roman emperor Constantine) and established there a center of learning which in the Middle Ages spread its influence into Europe. There was little interest in geometry at Byzantium however, and no progress in the science was made there.

Many of the Greek scientists and mathematicians and the Jewish physicians went from Alexandria to Arabia, chiefly to the city of Bagdad, and became teachers and physicians to the Arabs. In this way geometry became known to the Arabs. The Arab conquerors of Alexandria also preserved some of the scientific works from the Alexandrian Library. Among these were Euclid's *Elements* and Ptolemy's *Almagest*, which were translated into Arabic and became the standard works on geometry, trigonometry, and astronomy.

The Hindus, like the Chinese, claim to be the most ancient people on the earth, and that therefore all the sciences, including mathematics, originated with them. This is certainly not true of geometry but they have contributed to the development of arithmetic and algebra. The present-day so-called *Arabic Numerals* (1, 2, 3, 4, 5, 6, 7, 8, 9, 0) originated with the Hindus from whom they passed to the Arabs, who introduced them into Europe. (See the ARITHMETIC of this series.)

One of the earliest Hindu writers of any note of whom there is any definite record was Arya-Bhata, who was born in 476 A.D., the year of the fall of the old Roman Empire. Arya-Bhata was an astronomer and wrote on geometry as applied to astronomy. He is noted for his determination of the approximate value of the circumference-diameter ratio of the circle which is now used, viz., 3.1416. One of the best known of the Hindu mathematicians was Bhaskara, who was born in 1114 A.D. He wrote works on arithmetic, algebra, geometry and astronomy and is noted for an original proof of the Pythagorean Theorem.

The knowledge of arithmetic and algebra obtained by the Arabs from the Hindus in India and the knowledge of geometry, trigonometry and astronomy gained by them from the Greeks in Alexandria and Bagdad were all assiduously cultivated by them and among the Arabs arose the most learned scientists of the world for some six hundred years following the fall of Alexandria (during about 600–1200 A.D.). The Arab scholars did not contribute greatly to the development of geometry and trigonometry but they were very active and successful in the application of these sciences to astronomy. Their greatest work, their greatest contribution to the cause of science and the history of geometry, was the preservation of scientific knowledge during the Dark Ages, when the light of learning almost went out in Europe. They translated and published the learned works in science and established schools over all the great Arabian Empire which spread as far into Europe as Spain. Through them the science of geometry was finally introduced into Europe at the time of the great revival of learning.

7. Geometry and the Universities in the Middle Ages and Since. During the so-called Dark Ages in Europe, the period between the fall of Rome (476 A.D.) and the Revival of Learning (about 1200–1300 A.D.) the greater part of Europe was in almost total ignorance of the arts and sciences. Knowledge of the arts and of literature and philosophy was kept alive in the East at Byzantium and to a certain extent by the monks and theologians in the monasteries of Europe, while mathematics and astronomy were, as we have seen, preserved largely through the Arabs.

During the 12th century some of the monasteries formed schools which by the end of the 14th century had developed into the present great universities of Europe. At first the chief subjects taught in these schools were grammar, rhetoric, theology, and philosophy. Later medicine and the law were added and finally, during the 13th

Art. 7 ORIGIN AND DEVELOPMENT OF GEOMETRY 13

and 14th centuries, mathematics and the sciences. These latter came largely through the contact of Europe with the Arabs. Chief among the mathematical and scientific studies were arithmetic, geometry, trigonometry, and astronomy, with geometry receiving special attention. It was during these times that the Arabic and the Greek editions of Euclid's *Elements of Geometry* were translated into Latin and the European languages. One of the earliest of these translators was an English monk, Athelbard of Bath, who translated the *Elements* from Arabic into Latin in 1120 A.D. Another famous edition was that of the Italian monk Giovanni Campano, or as he is better known by the Latin form of his name, Johannes Campanus, who in 1250 A.D. reissued Athelbard's translation with improvements and a commentary. This edition was printed in Venice in 1482, the first book on geometry to be printed. The first translation of the *Elements* into English was made by Sir Henry Billingsley, Lord Mayor of London, in 1570. One of the most exactly correct English editions of the *Elements* is the translation made by John Williamson at Oxford in 1781.

After the entry of geometry into the universities it soon became a recognized part of a liberal education and was taught in the lower schools, but no extension or improvement of the science itself was made until modern times. The greatest advance was in the application and use of geometry in the arts and trades and in engineering. The only textbook in geometry used in England until very recent years was Euclid's *Elements of Geometry*. A very good edition suitable for school use is Todhunter's *Elements of Euclid* (published by Macmillan and Company, London and New York, and now also included in Everyman's Library).

In the attempts to prove the *parallel axiom* of Euclid (article 9) the entire system of Euclid's geometry was re-stated, simplified, and in some respects improved, by the celebrated French mathematician Adrien M. Legendre (1752–1833) who in 1794 published his results in a famous textbook, the *Eléments de Géométrie*. This work includes also, in addition to Euclid's original work, the work of Archimedes on the circle and the "three round bodies." This book soon replaced the original text of Euclid in France and the United States and for more than a hundred years has served as a standard and model. A widely used modern text, based on those of Euclid and Legendre, is Wentworth's *Plane and Solid Geometry* (published by Ginn and Company,

New York), which also contains a "book" or section on the conic sections (article 5).

8. Modern Development of Geometry. Since the introduction of geometry into Europe from the East at the time of the revival of learning and its adoption as an important part of a liberal education, most of the important advances in the science of geometry have been made in Europe. The chief of these will be mentioned, together with the important names and dates connected with these developments.

Among the earliest of the Europeans to consider the logical foundations of geometry was the Frenchman Gerard Desargues (1593–1662) who extended Euclid's ideas and methods to discover new geometrical principles and analyzed the fundamental principles of the older Euclidean geometry. He is thus one of the founders of what is known as Modern Pure Geometry. This field was further developed by several workers and is actively cultivated at the present time. The present trend is in the direction of what has come to be known as Projective Geometry, which deals with form and position without measurement or consideration of size. The subject was first fully formulated by a French military engineer, Jean Victor Poncelet (1788–1867) about the year 1822. Another Frenchman, Gaspard Monge (1746–1818), had developed about 1767, while a student in a school of drawing and surveying, a geometrical method of making drawings which he later further developed into the science or art known as Descriptive Geometry. It was at first preserved by the French military authorities as a military secret, but is now a large and important part of both geometry and engineering.

An entirely different development of geometry took place on the publication in 1637 of *La Géométrie* by the French philosopher and mathematician Rene Descartes (1596–1650), who was also a soldier. Descartes showed how geometrical figures could be described and their properties analyzed by means of algebraic equations. This method is applicable not only to the straight line and circle of elementary pure geometry but to any line, curve or figure whatever. It is called Analytical Geometry and is now one of the large and important branches of geometry. (See the chapters on GRAPHS in the ARITHMETIC and the CALCULUS of this series.) After Descartes showed how geometry can be treated algebraically other mathematicians began to apply the newly invented *calculus* (invented by Newton and Leibnitz about 1666–1676), to the study of geometrical curves and later to surfaces, either of which may be of any shape whatever. This application has developed into a branch of geometry known as Differential Geometry which is of the highest importance in modern science.

9. Non-Euclidean Geometry. All the developments discussed in this article have been based on Euclid's original principles and methods, and their extensions and applications, and the whole body of geometry composed of these various branches is known as *Euclidean* geometry. In the last hundred years however, there has grown up an entirely new idea of geometry which starts from a distinctly different viewpoint and is known as *Non-euclidean* geometry. This development is briefly outlined in Chapter 11.

In the first part of the present book only Euclidean geometry will be treated. A general textbook on geometry, based on those of Euclid and Legendre, and containing introductions to the modern (pure) geometry and to Non-euclidean geometry, is Phillips and Fisher's *Elements of Geometry* (published by American Book Company).

Chapter 2

SOME PRELIMINARY IDEAS AND METHODS OF GEOMETRY

10. What is Geometry? The ordinary ideas conveyed by the words "straight," "curved," "flat," "square," "round," "long," "wide," "thick," "space," and the like are familiar to everyone. The relations between such ideas as these and their application to the description and measurement of objects of everyday experience and observation are matters of common speech and knowledge, and unless attention is called to them in some peculiar manner or it is required to perform some simple calculation based on them it is hardly realized that these relations have any special significance. If a beautiful building or a complicated machine is carefully observed however, and an attempt is made to describe it in detail, or if it is required to solve one of the simple problems given in the chapters on mensuration (measurement) in arithmetic, it will be realized that a special significance attaches to this body of ideas and knowledge and that it is of the highest importance in our life and thought and in the industrial world.

When each idea of the set of ideas mentioned is clearly stated and the relations between them are analyzed, and when their logical consequences are followed out according to definite stated rules, there results a complete and clearly defined system or body of knowledge. This branch of knowledge is called *geometry*. Since this knowledge is systematically classified and all results obtained in it are subjected to logical processes, it is called a *science*.

Summarizing the last two paragraphs, we may say then that *geometry is the science which deals with the properties of space and their relations*. Here we have not defined "space." It is one of the ideas or notions which is not defined. From the viewpoint of its uses and applications we also say that *geometry is the art and science of description and measurement in space*. Neither of these definitions, or rather, characterizations, really *defines* geometry and probably would not satisfy the philosophers. For the purposes of this book, however, they will serve.

ART. 11   SOME PRELIMINARY IDEAS AND METHODS   17

In modern times, mathematicians have adopted a new attitude toward the concept of a mathematical science. First, there are identified a set of undefined terms. The most important undefined terms in geometry are point, line, and plane.

Let us consider the reason for basing a mathematical science upon a set of undefined terms. For example, let us assume that we wish to define "point." In order to do this we will have to use some words other than "point." In turn, it will be necessary to define one or more of the words used in defining "point." This process will have to be continued *ad infinitum*. In order to avoid such an infinite and unproductive chain of definitions we agree to start with a basic set of undefined terms. Although such terms are undefined we may have concrete associations which give us pictures of these terms. For example, a point is suggested by the tip of a needle, or by a distant star. A line is suggested by the edge of this page, or by a foul line on a baseball diamond. A plane is suggested by the top of a table, or by a wall in a room.

Thus, the undefined terms are abstract ideas although they may have concrete representations. For example, a point may be represented by a dot on this sheet of paper but this representation is not an actual point since a point has no dimensions. Similarly, a line may be represented by a pencil line but again, this is not an actual representation since a line has no width. Yet, the results obtained by reasoning with these undefined terms may be applied to the physical world in which we live. It should be observed that all other terms used in geometry such as triangle, right angle, vertex, etc., may be defined by using our undefined terms.

A similar situation exists in connection with the logical organization of geometry. In a logical chain of reasoning it is necessary to start with some basic principles. The truth of these principles cannot be demonstrated since there are no principles that can be used for support or proof. These starting principles are called "postulates." By a process of reasoning these principles, or postulates, may be used to generate and establish new principles. These derived principles are called "theorems." The body of this volume will be used to establish the theorems which are the main subject of geometry.

11. Divisions of Geometry. We learn early in our life experience to distinguish between an object which has "body" and occupies space, such as a block or a box, and the flat smooth surface of such an object, which does not occupy space but does have size and extent. The complete object is called a *solid* while the flat surface without thickness is called a *plane*. The body or object may be "solid" in the ordinary sense of a block of stone, or it may be "hollow," as an empty box. If it occupies a portion of space however, it is in either case called a *solid*. If a flat surface such as the smooth top of a table can be thought of as existing apart from the table, so that it has no thickness, and if there is no bend or irregularity in it, then it becomes the *plane* of geometry. (The word "plane" is not to be confused with "plain.")

All the lines and figures which can be drawn or formed on a plane are called *plane figures*, and that portion of geometry which treats of the construction, relations, description and measurement of plane figures is called PLANE GEOMETRY.

All the lines, figures and objects which can be drawn or formed in space without restriction to a plane are called solid figures or simply *solids*, and that portion of geometry which treats of the construction, relations, description and measurement of solids is called SOLID GEOMETRY.

This book will deal mainly with plane geometry, including only a chapter or so on solid geometry.

12. Methods of Euclidean Geometry. Plane and solid geometry as defined in the preceding article include consideration of every conceivable form of figure and object. The complete treatment of all these requires many different methods, most of which have been mentioned in articles 8 and 9. In the geometry originally formulated by Euclid (Chapter 1) the figures and constructions considered are restricted to those possible with the use of only the straight line and circle. This geometry is (as we have seen) called *Euclidean elementary geometry*.

Straight lines are drawn on (or in) a plane by means of a "straightedge" or *ruler*, and circles or parts of circles by means of the *compasses*, or as we shall say, the *compass*. The method of drawing or constructing figures and solving problems in elementary Euclidean geometry is therefore referred to as the method of "straight line and circle" or of "ruler and compass."

This book will deal only with elementary geometry.

13. Straight Line and Ruler. The notion of a *straight line* is fundamental and is difficult to define in simpler terms. Many definitions have been given, from Euclid to the latest writer. Few of these

Art. 13 SOME PRELIMINARY IDEAS AND METHODS

are entirely satisfactory, however. Similarly, the notion of a *point* is difficult to define. We shall consider *point* and *line* undefined terms. In drawing and representing a line of any kind it is referred to by naming two or more letters which indicate points on the line. Thus, in Fig. 1, (*a*) represents the straight line *AB*, and (*b*) represents the broken line *ABC*. Fig. 2 represents the curved line *ABCD*.

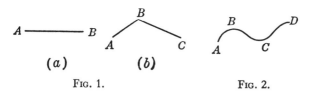

Fig. 1.   Fig. 2.

A *ruler* is any material object having an edge which is straight and which may be laid on a plane surface so that a pen, pencil or other marker may be held against that edge and drawn along the surface, thus marking off a straight line on the surface. Of course, the ideal straight line has no width, while the lines drawn by pens, pencils, crayons, etc., do have width. In all geometrical work these lines should be made as narrow or as "fine" as possible. A ruler may or may not have its edge divided and marked off in sections for measurement. A simple form of ruler which should be familiar is shown in Fig. 3, the edge *AB* being used as the "straight-edge."

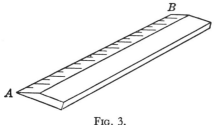

Fig. 3.

Another form of ruler is shown in Fig. 4. This is called a *T-square*. It may be moved up and down on the drawing board by sliding the straight edge of the "head" against the straight edge of the board. The straight edge of the "blade" is then used as a ruler to draw straight lines across the board as shown in the figure.

The device shown at *ABC* is made of wood or cut from one piece of celluloid or similar material, and has three straight edges AB, BC, CA. This device is a form of ruler, called a *triangle*. The edge AC or BC of the triangle may be held against the blade of the T-square and moved along it, while the other two edges of the triangle are used to draw "up and down" or inclined lines, or AB may be held against the blade and both edges AC and BC used to draw inclined lines.

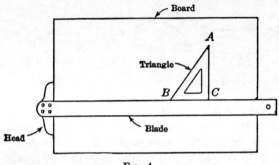

Fig. 4.

In order to do careful and correct geometrical drawing the student should have all the articles shown in Figs. 3 and 4, together with a pen and a pencil with which fine lines can be drawn, and thin-headed "thumb tacks" with which paper is easily fastened to the drawing board.

14. Circle and Compass. The familiar "round" plane figure shown in Fig. 5 is called a *circle*. Strictly speaking, the curved line

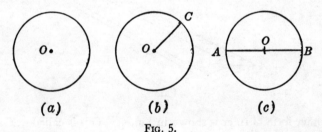

Fig. 5.

forming the figure is itself the circle. This line is sometimes called the *circumference* of the circle, though, strictly speaking, the word "circumference" should be reserved to designate the measure of the *length* of this curved line.

ART. 14   SOME PRELIMINARY IDEAS AND METHODS

Of all the points which lie inside the circle but not on the curved line itself, there is one which is at the same distance from every point on the circle. This point is called the *center* of the circle, and is indicated at O in Fig. 5(a).

A straight line joining the center with any point on the circle, as the line OC in Fig. 5(b), is called a *radius* of the circle, and the length of this line is referred to as *the radius*.

Any straight line passing through the center and having its two ends on the circle, as AB in Fig. 5(c), is called a *diameter* of the circle, and the length of this line is referred to as *the diameter*.

It is seen at once that the diameter is twice the radius, or the radius is half the diameter.

It will be recalled from arithmetic that in any circle the circumference is equal to the diameter multiplied by a certain number. In a later chapter this will be definitely proved and the value of this multiplying number will be calculated.

The *compass* (or as sometimes called, *compasses*) is an instrument used for drawing circles. The compass and its use are shown in Fig. 6. The handle H and the *legs* JC and JP are usually made of metal and the three are joined by a pivot at J. C is a sharp point which may be lightly pressed into the paper so as to be held stationary, and P is a pen or pencil point fastened to one leg.

To draw a circle with the compass the point C is lightly pressed into the paper and P allowed to rest on the paper, with the handle H held between the thumb and one finger. The compass is then inclined slightly and the handle rolled between the thumb and finger. With C held stationary P will then trace out the circle, as shown by the curved line in Fig. 6, the arrow heads indicating the direction of motion of the point P.

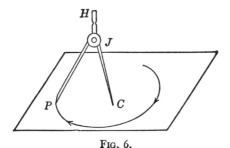

FIG. 6.

The position of C is the center of the circle, and the opening or *spread* of the legs, PC, is the radius of the circle.

The compass is a necessary part of the equipment for geometrical drawing and should be added by the student to the equipment shown in Figs. 3 and 4.

Compasses having two sharp points instead of one point and a pencil or pen, are called *dividers*. By setting the two sharp points on any two points on a drawing or a graduated ruler, the distance between them, or the length of any section of a line, may be marked off, or *laid off*, as it is said, with the dividers. Whenever a distance is spoken of as being laid off equal to another distance it is understood to be done with the dividers.

By setting the points of the dividers at any chosen distance this distance or length may be *stepped off* on a line by swinging each point in turn around the other as a center, thus dividing the line into equal sections or *segments*. It is from this operation that the name "dividers" came.

15. Some Fundamental Definitions. With the foundation ideas stated above as a basis, we can now state some of the fundamental definitions of geometry. Some of the statements already made will be partially repeated for the sake of completeness and systematic procedure.

A *broken* line is a line formed of successive sections, or *segments*, of straight lines.

A *curved* line, or simply a *curve*, is a line no portion of which is straight.

The forms of lines defined or described here are already shown in Fig. 1 and Fig. 2, article 13.

A *plane surface* is such a surface that the straight line joining any two points in it lies entirely in the surface. This means that if a straight line is laid on a plane surface in any position or direction it will touch the surface all along its length. This is expressed in ordinary language by saying that a plane surface is one which is perfectly "flat." A *plane surface* is usually called simply a *plane*.

A *curved surface* is a surface no portion of which is plane.

Solids are classified according to the nature of their bounding surfaces, as we shall consider later in more detail.

A *geometrical figure* is any combination of points, lines, surfaces and solids, formed under specified conditions.

Figures formed by points and straight lines or curved lines all of

Art. 16 SOME PRELIMINARY IDEAS AND METHODS

which lie in a plane are called *plane figures*. Plane figures have not more than two dimensions.

Solid figures, or *solids*, have already been defined; it is now seen that a solid figure may also be thought of as one which does not lie wholly in one plane.

Geometry has also been already defined or characterized. We can now say more definitely, however, that

GEOMETRY *is the science of extension and position in space, or the science of position, form and magnitude.*

More specifically, geometry may be thought of as the science which treats of the *construction* of figures under given conditions and of their *measurement* and their *properties*. Using the second form of statement, then,

PLANE GEOMETRY *is the science which treats of plane figures.* It is geometry of *two dimensions*.

SOLID GEOMETRY *is the science which treats of figures in space.* It is geometry of *three dimensions*.

16. General Terms Used in Geometry. We collect here some of the terms generally used in geometry, together with their concise definitions. It is believed that they will be understood without extended discussion or explanation. It is not necessary to commit these definitions to memory, as they will be used and illustrated frequently.

In geometry, as elsewhere, a *proof* is a process of reasoning by which the truth or falsity of any statement is logically determined and established.

A *theorem* is a statement to be proved.

A *postulate* is a statement admitted to be true without proof.

A *construction* is the representation of a required figure by means of points and lines.

A *problem* is a construction to be made so that it shall satisfy certain conditions, or a measurement or calculation to be made so as to obtain a certain required result or results. When in either case the requirement is satisfied, the problem is said to be *solved*.

The term *proposition* is sometimes used to indicate a theorem, axiom, postulate or problem, that is, anything proposed for proof, performance or discussion. In this book a proposition will mean a theorem or problem.

The full statement of a theorem consists of two parts: the *hypothesis*, that which is granted or assumed; and the *conclusion*, that which is asserted to follow from the hypothesis.

The theorems and problems of geometry form the most important part of its content or subject-matter. After the various definitions are given, so that geometry may have a vocabulary, and the postulates are stated, so that geometry may have a starting point, all the substance or content of geometric truth or geometry proper is contained in the theorems and problems. When the theorems are proved and the problems solved they contain all geometrical knowledge.

(The term "theorem" must not be confused with "theory." A theory may or may not be capable of proof, while a theorem is always capable of proof.)

The statement and proof of the theorems and the statement and solution of the problems are therefore the chief immediate objects of geometry.

17. Theorems and Their Proof. In the preceding article a *theorem*, consisting of the hypothesis and the conclusion, has been defined as a statement to be proved. Such a statement as

If two straight lines intersect, the opposite angles so formed are equal

is a theorem. The first part of the statement, preceding the comma, contains the hypothesis, or given information, and the second part the conclusion. It is this second part of the complete statement which is to be proved.

The form in which a proof is stated or written may be one of several. Thus the theorem may be stated, a figure drawn to represent the hypothesis and marked with letters for convenient reference, and the hypothesis and conclusion then expressed in terms of these letters. Beginning with the hypothesis or an additional construction if necessary, the various pertinent steps in the argument are then made in proper and logical sequence and the reason given for each, until finally one of these statements is the desired conclusion itself. The theorem has then been proved.

This is the method used by Euclid and has been very widely used since his time. The statements are usually very careful and precise and as short as possible, and the whole is very formal and exact. Nothing necessary is omitted and nothing unnecessary is included. In the strict Euclidean form of proof the letters Q.E.D. are added

ART. 18   SOME PRELIMINARY IDEAS AND METHODS

after the final conclusion. They stand for the Latin words *quod erat demonstrandum*, which mean "which was to be shown." The Euclidean method is frequently used in school textbooks because of its conciseness and its convenience for reference, and on account of the fact that a school course in geometry is intended to be a training in mental discipline as much as in the useful properties of geometrical figures. This method is used in the books of Todhunter, Wentworth, and Phillips and Fisher mentioned in articles 7 and 9.

Another form of proof is one which begins with a general discussion of a previously constructed figure, and in an informal manner, as in conversation or description, brings out successive facts or properties, using additional construction wherever necessary, until an important conclusion is reached. The full theorem is then stated for the first time, containing the conclusion just reached. This method—also very old—is easier to follow than that of Euclid but is not so convenient for reference. It was used very effectively by Legendre in the book mentioned in article 7, and is still widely used in modern research work and in applications of geometry.

In this book both methods will be used, together with combinations of the two, and in general the proofs will be made as simple and informal as possible consistent with interest and correctness.

18. Problems and Their Solutions. A *problem* is distinguished from a theorem by reason of its being something which is proposed to be done rather than a statement made to be proved. The thing required to be done may be the construction of a geometrical figure which shall satisfy certain conditions as, for example,

To construct a triangle, having given the three sides,

or it may be the derivation of a formula, as in algebra or in the chapters on geometrical measurement in arithmetic, or it may be the calculation of a required number, as in ordinary arithmetic or algebra.

If the problem is a geometrical construction the various parts of the figure are constructed as required, based on the postulates, and each step in the construction is stated or described as it is made. When the final result is obtained it is stated to be the result required, and this statement is then proved to be correct, after the manner of proving a theorem. In the Euclidean method this procedure is followed strictly and formally and at the end are placed the letters Q.E.F., meaning "which was to be done" (*quod erat faciendum*).

Modifications of this procedure will be illustrated as we go along, and the triangle problem, stated above, will be solved. If the problem requires the derivation of a formula or the calculation of a numerical result, the derivation or calculation is itself the proof and no separate formal proof is required.

19. Starting Points of Geometry. Every discussion, every study of any particular subject, must have some starting point. There must be some truth or fact which is known or some statement which is assumed, admitted, or taken for granted. When these statements are once allowed they are not thereafter to be changed or denied, and every result or conclusion drawn from them by a correct logical process must be admitted as also true, or at least as true as the original assumption or statements.

If a different set of statements is taken as basis, it is to be expected that different results will be obtained. In geometry the selection of these fundamental statements has been guided by four considerations: (1) They must be such as will offer no contradiction to our common knowledge and experience; (2) they must be mutually consistent; (3) all results and conclusions obtained from them must be consistent with one another; (4) they should be as few as possible. Thousands of years of examination, study and test have shown the original basic statements selected by Euclid and his forerunners to be generally satisfactory and for the ordinary purposes of elementary geometry they have not been improved upon.

The basic, fundamental statements of geometry are its *postulates*, already defined in article 16. Since these are, so to speak, the foundation stones of the structure they must obviously be laid upon "bedrock," so deep down that it is never necessary or desirable to go lower. That is, the postulates must be stated in terms of words already defined or generally understood, they must be put in as simple form as possible, and it should be unnecessary to analyze or explain them further after they are stated.

The postulates of geometry are of two kinds: those having to do with common knowledge; and those having to do particularly with the things especially considered in geometry, such as those things defined in articles 16 and 17. The postulates of the first kind were called by Euclid "common notions." The postulates of the second kind are the *geometrical postulates*.

ART. 20 SOME PRELIMINARY IDEAS AND METHODS 27

The general postulates are collected together for reference in the next article and the geometrical postulates in the article following.

20. General Postulates. The general postulates, which apply not only to geometry but to all mathematics, are here stated in a form suitable for convenient use, and without discussion.

(1) *Things which are equal to the same thing or to equal things are equal to each other.*

(2) *If equals are added to equals the sums are equal.*

(3) *If equals are taken from equals the remainders are equal.*

(4) *If equals are added to or taken from unequals the results are unequal in the same order as at first.*

(5) *The doubles or any equal multiples of equals are equal; and those of unequals are unequal in the same order as at first.*

(6) *The halves or any equal parts of equals are equal; and those of unequals are unequal in the same order as at first.*

(7) *Equal powers and roots of equals are equal.*

(8) *The whole is greater than any of its parts.*

(9) *The whole is equal to the sum of all its parts.*

(10) *In any mathematical operation any thing may be substituted in the place of its equal.*

Parts of some of these postulates may be equivalent to other postulates but it is better for our purposes not to combine them, but to use them as here stated.

Postulate (1) is frequently referred to as the *equality postulate.*

Postulates (2) and (3) are referred to as the *addition* and *subtraction postulates,* respectively.

Postulate (4) is sometimes called the *inequality postulate.*

Postulates (5) and (6) are frequently referred to as the *multiplication* and *division postulates,* respectively.

Postulate (7) is similarly the *power and root postulate.*

Postulates (8) and (9) may be called the *postulates of the whole.*

Postulate (10) is the *substitution postulate.*

In this book the general postulates will be referred to by the names just given or by the numbers with which they are listed above.

To the reader who has studied algebra the general postulates will be familiar as having been used in studying the properties of equations.

21. The Geometrical Postulates. Although the geometrical postulates will be better understood later when we have given the definitions and discussions of certain of the terms involved, they are collected and stated here for the sake of completeness and for convenient reference.

(11) *Through any two points in space there can be one and only one straight line.*

(12) *A straight line may be produced to any length.*

(13) *A circle may be described with any given point as center and with any given radius.*

(14) *Motion of a geometrical figure does not change its size or shape.*

(15) *All right angles are equal.*

(16) *One and only one parallel to a given straight line can be drawn through a point outside the line.*

Postulate (11) is called the *straight line postulate*.

Postulate (14) is called the *superposition postulate*.

Postulate (16) is the famous *parallel postulate*.

Part II

PROPERTIES OF PLANE FIGURES

Chapter 3

SOME PROPERTIES OF STRAIGHT LINES AND ANGLES

22. Straight Lines. A *straight line* has already been defined as *the shortest line joining two points*. When this definition is once stated the straight line postulate (*through any two points there is one and only one straight line*) is obvious. Two lines which have only one point in common are said to *intersect* at that point, and are called *intersecting* lines.

From the straight line postulate the following two statements are at once seen to be true:

I. *Two straight lines passing through the same two points coincide and form but one line.*
II. *Two straight lines can intersect in only one point.*

For, if they had two points in common they would coincide and be but one line.

According to the postulate which states that a straight line may be extended indefinitely, any straight line may be thought of as being of indefinite length.

The straight line passing through the points, A, B, Fig. 7, may extend indefinitely on either side of these points. The portion of the straight line lying between A and B is called a *segment* of the line; thus "the segment AB."

FIG. 7.

It is usual to refer to a straight line simply as a *line*, and the single word "line" will hereafter be understood as meaning a *straight line*.

The length of a line segment joining two points is called the *distance* between the two points. Thus in Fig. 7 we refer to the distance AB.

If two line segments have the same length they are said to be *equal*.

Distances, or line lengths, may obviously be added or multiplied to give lines of greater length, and also subtracted or divided to give lines of less length.

If a line is divided into *two equal* parts it is said to be *bisected*. If one line crosses another at the point of bisection (middle point) the first line is called a *bisector* of the second.

23. Angles. When two straight lines meet at a point or are drawn from the same point in different directions, the figure which they form is called an *angle*. The two lines are called the *sides* of the angle and their meeting point is called the *vertex* of the angle. Thus in Fig. 8 the two lines AB and AC form an angle. AB and AC are the sides of the angle and A as its vertex.

The angle in Fig. 8 is referred to as the "angle at A" or by naming with capital letters the vertex and a point on each leg, as the "angle BAC," the letter representing the vertex being between the other two. For brevity this is often written as ∠BAC, the symbol ∠ replacing the word "angle." The plural, "angles," is indicated by the symbol ∡. Similarly to Fig. 8 we have in Fig. 9 ∠EDF.

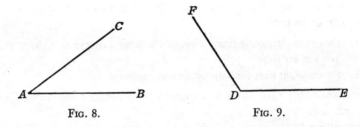

Fig. 8. Fig. 9.

If there is but one angle at a given vertex, as in Figs. 8, 9, the angle may be designated by the letter at the vertex alone, as ∠A, ∠D. If there are several angles at the same vertex however, as in Figs. 10, 11, this designation is not sufficient and three letters must be stated in naming each angle. Thus in Fig. 10 we have the three angles BAC, BAD, CAD.

Another method used in a case like this is shown in Fig. 11. Here ∠AOB is referred to as ∠a, ∠BOC as ∠b, and ∠COD as ∠c. The short curved line joining the sides of the angle may be used as for ∠a, ∠b, or not, as at ∠c.

The size or measure of an angle is determined by the *amount of opening* between the sides, and NOT by the lengths of the sides; that is, by the *sharpness* or *bluntness* of the "corner" formed by the sides. Thus, in Figs. 8, 9, ∠EDF is greater than ∠BAC; in Fig. 10 ∠BAC is less than ∠BAD; in Fig. 11 ∠b is greater than ∠a, and ∠a is less than ∠c.

Two angles are said to be *equal* if they can be placed together so that their vertexes (or vertices) are at the same point and the two sides of one coincide with the two sides of the other. This is a very important definition.

When several lines meet at one point to form more than one angle, any two of the angles which have one side in common are said to be *adjacent*. Thus in Fig. 10 $\angle BAC$ and $\angle DAC$ are adjacent, and similarly in Fig. 11 $\angle a$, $\angle b$ and $\angle b$, $\angle c$ are adjacent pairs.

When a line is drawn through the vertex of an angle between the sides it is said to *divide* the angle. Thus in Fig. 10 the line AC divides the angle BAD into the two angles BAC, DAC, and in Fig. 11 the lines OB, OC divide the angle AOD into the three angles a, b, c.

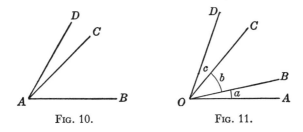

Fig. 10.   Fig. 11.

If a line divides an angle into *two equal* parts it is said to *bisect* the angle, and is called the *bisector* of the angle. If two lines divide an angle into *three equal* parts they are said to *trisect* the angle. Thus in Fig. 12 $\angle a = \angle b$ and AC is the bisector of $\angle BAD$; in Fig. 13 $\angle x = \angle y = \angle z$ and the lines AC, AD trisect $\angle BAE$.

The last two paragraphs show how angles may be *divided;* thus in Fig. 12, $\angle a = \angle BAD \div 2$ or $\frac{1}{2}\angle BAD$, and in Fig. 13, $\angle x = \angle BAE \div 3$ or $\frac{1}{3}\angle BAE$. Similarly angles may be *multiplied*. Thus in Fig. 12, 13, $\angle BAD = 2 \times \angle b$ or $2\angle b$, $\angle BAE = 3 \times \angle x$ or $3\angle x$.

Angles may also be *added* and *subtracted*. Thus in Figs. 10–13 the

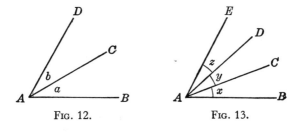

Fig. 12.   Fig. 13.

whole angle in each case is the sum of the several parts, and any one of the parts may be subtracted from the whole to give as remainder the other part or sum of the other parts.

24. Perpendicular Lines and Right Angles. Two intersecting straight lines form four angles which have the same vertex, as in Fig. 14, where the two lines *AB*, *CD*, intersecting at *P*, form the four angles *APC*, *CPB*, *BPD*, *DPA*.

If the four angles formed by two intersecting lines are all equal, the lines are said to be *perpendicular* to each other, and each of the four equal angles is called a *right angle*. Thus, in Fig. 14 the lines *AB*, *CD* are perpendicular to each other and each of the angles *BPC*, *CPA*, *APD*, *DPB* is a right angle. A right angle may be thought of as one whose sides form a "square corner."

It is to be noted that in Fig. 14 neither line *alone* is said to be perpendicular, but only with reference *to the other*. Thus *CD* alone is not perpendicular, it is perpendicular *to AB*. Similarly *AB* is perpendicular *to CD*. The term "perpendicular" must not be confused with "vertical." Thus the wall of a building is *vertical* but it may not be *perpendicular to the* ground. It is perpendicular to the *ground* if the ground is level, or horizontal, but the wall cannot be referred to simply as "a perpendicular wall." In this case the ground is also perpendicular, to the wall.

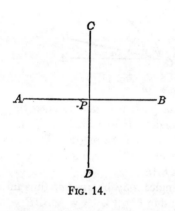

Fig. 14.

Similarly a line cannot without further reference be referred to simply as "a perpendicular line." If, however, the line *to which* it is perpendicular is given, then the two *together* are called *perpendicular lines* (plural) or simply *perpendiculars*. Similarly *with reference to the other line of the pair* either line is referred to as *a perpendicular*. Thus in Fig. 14, with reference to line *CD*, *AB* is a perpendicular; with reference to *AB* the segment *CP* or *DP* is referred to as the perpendicular *at P*. The sign ⊥ is used for the word "perpendicular."

Furthermore, two lines need not be vertical and horizontal to be perpendicular to each other; they may be together inclined in any position so long as they form four equal (right) angles.

ART. 25 PROPERTIES OF STRAIGHT LINES AND ANGLES 35

From the definition of a right angle given at the beginning of this article the postulate *all right angles are equal* is now understood and is indeed self-evident.

As an independent result we now have also the following property of a pair of perpendiculars:

III. *At a given point in a line there can be but one perpendicular to the line.*

For, if there could be more than one perpendicular they would be distinct lines forming different angles and we should have right angles of different sizes; but this is impossible.

25. Other Angles and Their Relations. An angle which is greater than a right angle but less than two right angles is called an *obtuse* (dull) angle. An angle which is less than a right angle is called an *acute* (sharp) angle.

Two angles whose sum is equal to a right angle are said to be *complementary* angles, and each is called the *complement* of the other.

Two angles whose sum is equal to the sum of two right angles are said to be *supplementary* angles, and each is called the *supplement* of the other.

In Fig. 15 ∠BAD is a right angle and angles BAC, CAD are complementary; ∠BAC is the complement of ∠CAD, and vice versa.

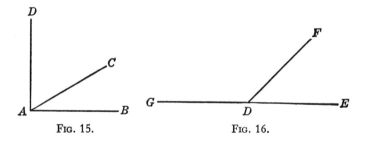

Fig. 15.   Fig. 16.

In Fig. 14 AB is a straight line and the sum of the two right angle BPC and CPA includes all possible angles formed on one side of the line AB. Similarly in Fig. 16 GE is a straight line and the sum of the two angles EDF and FDG is the same as the sum of two right angles. Thus the two angles EDF, FDG are supplementary; ∠EDF is the supplement of ∠FDG, and vice versa.

In Fig. 14 any pair of adjacent right angles, as angles BPC, CPA or angles CPA, APD are supplementary, for their sum is of course two right angles. But either pair of these right angles is formed by the perpendicular at P and the straight line to which it is perpendicular, as CP and the line AB. Similarly the sides DE and DG (called the *exterior* sides) of the adjacent supplementary angles EDF, FDG in Fig. 16 form the straight line GE. These considerations show that in general

IV. *If two adjacent angles are supplementary, their exterior sides are in the same straight line;* and also

V. *If two adjacent angles have their exterior sides in a straight line, the angles are supplementary.*

These two propositions might be formally proved but they are already seen to be true.

Since the sum of any two (or more) angles is, of course, also an angle, as in Figs. 10, 15, for example, so in Fig. 14 the sum of the two right angles BPC, CPA is the angle BPA. This single angle, *whose sides lie in a straight line*, is called a *straight angle*. So also in Fig. 16 the sum of the angles EDF, FDG is the straight angle EDG.

Thus *a straight angle is equal to two right angles*, or a right angle is equal to half a straight angle. Obviously *all straight angles are equal.*

26. Extension of the Meaning of Angles. Instead of thinking of an angle as being formed by two distinct lines, as AB and AC in Fig. 8, we may also think of it as being formed by the side AB turning about A as a pivot point, like the hand of a clock, until it reaches the position AC, as indicated by the curved arrow in Fig. 17. In this case the line AB is said to *turn through* or to *generate* the angle BAC. AB is the *initial* position and AC the *terminal* position of the generating line. Similarly, in Fig. 18 the line DE has generated $\angle EDF$ by rotating about the point D from the initial position DE to the terminal position DF.

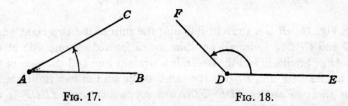

Fig. 17.   Fig. 18.

From this viewpoint the various angles in Fig. 19 are generated by the rotation of the line OB about O from the initial position OB to the successive terminal positions OE, OC, OF, OA, OG, etc., as indicated by the various curved arrows.

Here the angle BOC is, as before, a right angle; when the rotating or generating line has reached the position OA it has turned through two right angles ($\measuredangle BOC$, COA) and generated the straight angle BOA; and when it has reached OD it has turned through three right angles and generated the angle BOD.

An angle greater than a straight angle, such as $\measuredangle BOD$, BOG, BOH, indicated by the arrows, Fig. 19, is sometimes called a *reflex* angle.

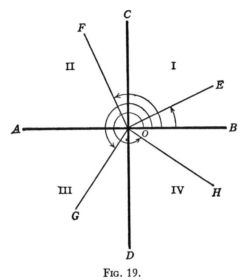

FIG. 19.

When the generating line has rotated entirely around the point O and returned to the initial position OB it has turned through four right angles or two straight angles, the total possible sum of all the angles about the point O. Thus

VI. *The whole angular magnitude about a point in a plane is four right angles, or two straight angles.*

Each of the four right angles at O, Fig. 19, is one fourth of the complete angle or circle, and is called a *quadrant* of the circle. It is customary to measure angles by rotation counter-clockwise (opposite to the rotation of the hands of a clock) from the horizontal line to the right of the vertex as initial line. The quadrants are then numbered in order as I, II, III, IV, as shown in Fig. 19. The angle BOE is then said to be in the *first* quadrant; $\angle BOF$ is then in the *second*, $\angle BOG$ in the *third*, and $\angle BOH$ in the *fourth*, quadrant.

From our previous definitions it is now seen that any acute angle is in the first quadrant and any obtuse angle in the second, while a reflex angle is in the third or fourth quadrant. The entire first quadrant represents a right angle and the entire first two quadrants represent a straight angle.

The method of designating angles explained in this article is of the greatest importance in trigonometry, mechanics and higher mathematics, and is the basis of the common system of angle measure, which we now discuss.

27. Measurement of Angles. Since the whole angular magnitude about a point or vertex is the complete angle or circle, the circle is the natural basis of angle measurement. The ancient Babylonians (article 1) are supposed to have divided the complete angle into 360 parts because in their time the heavens were supposed to complete the annual circuit about the earth in 360 days. On this account each of the 360 parts of the circle is called a step, grade or *degree*, the last name being the one most used. Therefore,

The unit of angle measurement is the 360th *part of a circle or complete angle, and is called a* DEGREE.

Since, according to the result VI in article 26, the complete angle contains four right angles, a right angle therefore contains one fourth of 360 or 90 degrees. On this account the degree angle is sometimes defined as one 90th of a right angle, and this is a convenient way to visualize an angle of one degree. Thus in Fig. 15 suppose that instead of the single line AC dividing the right angle BAD, there are 89 such lines dividing the right angle into 90 *equal* angles. Each of these very small, very acute angles is then an angle of one degree.

Since a right angle contains 90 degrees, a straight angle contains 180 degrees. An obtuse angle contains more than 90 but less than 180 degrees. A reflex angle then contains more than 180 degrees, and, as defined, the complete angle, contains 360 degrees. Similarly, as in Fig. 19, an angle of less than 90 degrees is in the first quadrant; one of between 90 and 180 degrees is in the second quadrant; one of between 180 and 270 in the third, and one of between 270 and 360 degrees in the fourth quadrant.

ART. 27 PROPERTIES OF STRAIGHT LINES AND ANGLES

If a right angle is bisected, each of the parts is an angle of 45 degrees; if trisected, each part is an angle of 30 degrees, and two of these adjacent parts form an angle of 60 degrees. These angles are much used and the draftsman's celluloid triangles (article 13) are usually cut of such shape as to have angles of 30, 45, 60, 90 degrees. When used to express the size of an angle, the word "degree" is usually replaced by the symbol (°), a small circle placed just above and to the right of the number. Thus *five degrees* is written 5°; a right angle is an angle of 90°; one tenth of a circle is 36°, which indicates the fraction 36/360 of the circle.

For measuring fractions of a degree, the degree is divided into 60 more minute parts and each is called the *minute degree* or simply the *minute*. The minute sign is the *prime* or "first" mark ('); thus half a degree is 30 minutes, written 30'. The minute is again divided into 60 parts, each of which is called the *second division* or simply *second*. The second mark is (''); thus one fourth of a minute is 15 seconds, written 15''. Using these subdivisions and symbols, angle sizes are then indicated by stating the number of degrees, minutes, and seconds which an angle contains (Ptolemy, article 5). Thus $42\frac{3}{8}$ degrees, or $42\frac{3}{8}°$, or 42.375°, is the same as 42° $22\frac{1}{2}'$, or 42° 22.5', or finally 42° 22' 30'', which is read "42 degrees, 22 minutes, 30 seconds."

Another system of angle measure, also based on the circle, is also used in higher mathematics and applications. This system is studied in trigonometry.

An instrument used to measure and draw angles is based on the circle and degree system of measurement. This is the *protractor*, shown in Fig. 20. It consists of a half circle cut from cardboard, celluloid or metal, with the inner part

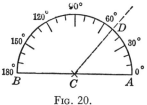

FIG. 20.

sometimes cut out so as to leave only a rim. The center C is marked on the straight edge AB which is the diameter of the circle, and the curved edge is divided up and marked off in degrees, from 0° to 180° in both directions (on both sides or faces). On the larger instruments the divisions show halves and smaller fractions of degrees.

To measure an angle, the protractor is placed on the figure with C at the vertex of the angle and CA along one side. The other side of the angle will then lie across the degree scale at a point which indicates the number of degrees in the angle. Thus in Fig. 20, $\angle ACD = 50°$.

To draw an angle of a stated number of degrees, the vertex point and one side are marked on the drawing and the protractor placed with C at the vertex and CA along that side. A mark is then made on the paper at the desired degree reading on the scale, as at D, the protractor is removed, and the line CD then drawn with a ruler. The angle ACD is then the desired angle.

28. Exercises.
1. What is the sum of 27° 18′ and 56° 47′?
2. What is the complement of 73° 42′?
3. What is the supplement of 95° 20′?
4. With a protractor lay off adjacent angles of 46° and 27°, as in Fig. 12, and measure their sum (the total angle).
5. Construct without a protractor angles of 45° and $22\frac{1}{2}°$.
6. Draw two supplementary angles, as in Fig. 16, and draw the bisector of each angle. Show that the bisectors are perpendicular to each other.
7. Do all the general axioms (article 20) hold good for negative and fractional numbers such as those met with in algebra and arithmetic?
8. The earth rotates on its axis once in 24 hours. Through how many degrees does it turn in one hour?
9. In a certain wheel there are 18 spokes. What is the angle between each two successive spokes?
10. Through what angle does the minute hand of a watch turn in 35 minutes of time?
11. Through what angle does the hour hand of a watch turn in 48 minutes of time?
12. Through what angle does the hour hand turn in a day?
13. Through how many degrees of angle does the second hand of a watch turn in one second of time?

29. Some Properties of Perpendicular and Oblique Lines.
Perpendicular lines have been defined as intersecting lines which make equal angles about the intersection point.

Intersecting lines which are not perpendicular are said to be *oblique*, that is, they are "inclined" to each other.

In contradistinction to oblique or "inclined" lines, lines which intersect perpendicularly are sometimes said to be "at right angles" to each other.

When two straight lines intersect, either perpendicularly or obliquely, forming four angles, the two angles which are opposite to each other are called *vertical* angles.

Thus in Fig. 14 the two angles BPC, APD are vertical angles, as are also the pair *BPD*, *APC*. Similarly, in Fig. 21 angles *a*, *b* are vertical angles, and also *x*, *y* are verticals. Each angle of a pair of vertical angles is called the vertical angle, or simply the *vertical*, of the other. Thus in Fig. 21, ∠*a* is the vertical of ∠*b*, and vice versa; and similarly for the vertical pair ∠*x*, *y*.

We can now prove the following interesting and important theorem:

VII. *If two straight lines intersect, the vertical angles are equal.*

Let Fig. 21 represent the two pairs of vertical angles formed by the intersecting straight lines *AB*, *CD*. We then have to show that ∠*a* = ∠*b*, and ∠*x* = ∠*y*.

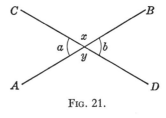

Fig. 21.

Consider the two adjacent angles *a*, *x* with their exterior sides in the straight line *AB*. Then, according to the proposition V, article 25, ∠*a* and ∠*x* are supplementary. That is, ∠*a* is the supplement of ∠*x*. Also adjacent angles *b*, *x* have their exterior sides in the straight line *CD*. Therefore, in the same way, ∠*b* is the supplement of ∠*x*. Thus, ∠*a* and ∠*b* are *each* the supplement of ∠*x*. They are thus equal to the same thing and (by Postulate 1, article 20) therefore equal to each other. That is, ∠*a* = ∠*b*, as was to be proved.

In the same manner we obtain the result ∠*x* = ∠*y*. The proposition is therefore established.

The preceding proof or demonstration is a simple illustration of the general method of geometry. That part of VII which precedes the comma is the hypothesis, and the remainder of the statement is the conclusion.

The property of vertical angles expressed in VII will later be found to be very useful.

In the statement III, article 24, we have already an important property of a perpendicular to a line. We shall now demonstrate two or three other such properties.

VIII. *Two straight lines drawn from a point in a perpendicular to a given straight line, and cutting off on the given line equal segments from the foot of the perpendicular, are equal in length and make equal angles with the perpendicular.*

In Fig. 22 let AB be the given line, CD the perpendicular to AB with its foot at D, and CE, CF the lines from point C in the perpendicular, cutting off the equal segments DE, DF on AB.

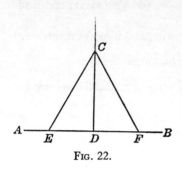

Fig. 22.

We have to prove $CE = CF$ and $\angle DCE = \angle DCF$.

Suppose CDA folded over, along CD, until it falls on CDB. Then DA will lie along DB, because the two right angles at D are equal, and by definition of equal angles their sides will coincide. Then also since by the hypothesis $ED = EF$, point E will fall on F.

Therefore the two ends of CE coincide with the two ends of CF, and by the straight line postulate they will coincide and be one line. Therefore, $CE = CF$, as was to be proved.

We then have also the vertices C, C and the lines CD, CD and also CE, CF coinciding, and therefore, by the definition of equal angles, $\angle DCE = \angle DCF$. —Q.E.D.

A perpendicular to a line at a point in the line is said to be *erected* at that point.

A perpendicular erected at the middle point of a given line is called the *perpendicular bisector* of the line.

IX. *Any point on the perpendicular bisector of a line is equally distant from the two ends of the line.*

In Fig. 23 let AB be the line, C its middle point, PC the perpendicular through C, and P any point on PC.

We then have to prove $AP = BP$.

Since by hypothesis $CA = CB$ and PC is the perpendicular at C, then by the proposition VIII already proved, we have at once that $AP = BP$. —Q.E.D.

X. *Only one perpendicular to a line can be drawn from a point outside the line.*

In Fig. 24 let AB be the given line with $PC \perp AB$ drawn from the external point P, and let PD be *any* other line from P to AB.

We then have to prove that PD is not perpendicular to AB.

As an added construction, for aid in the proof, extend the straight

Art. 29 PROPERTIES OF STRAIGHT LINES AND ANGLES

line PC to P' as shown dotted, making $P'C = PC$, and draw also $P'D$.

We then have that PCP' is a straight line, by the construction, and hence PDP' cannot be a straight line, by the straight line postulate (only one straight line between two points). Hence $\angle PDP'$ is not a straight angle.

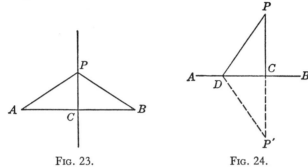

Fig. 23.　　　　　Fig. 24.

Again, since by hypothesis $DC \perp PP'$ and by construction $CP = CP'$, then by proposition VIII $\angle CDP = \angle CDP'$, and hence $\angle CDP$ is half of $\angle PDP'$.

But as already shown, $\angle PDP'$ is not a straight angle. Hence its half, $\angle CDP$ is not a right angle. Therefore PD is not perpendicular to AB. 　—Q.E.D.

This states that PC is the only perpendicular to AB from the external point P, because PD is *any* other line than PC.

XI. *The perpendicular is the shortest line that can be drawn to a straight line from an external point.*

In Fig. 24 let AB be the given straight line, P the external point, PC the perpendicular from P, and PD *any* other line from P to AB.

We then have to prove that PC is shorter than PD.

Suppose the additional lines $P'C$, $P'D$ constructed as already stated for Fig. 24.

Then since $P'C = PC$ and $AC \perp PP'$, as in VIII, $P'D = PD$. Hence, the broken line $PDP' = P'D + PD = 2 \cdot \overline{PD}$, and also $PP' = 2 \cdot \overline{PC}$ by construction.

Now of the two lines joining PP' the straight line PP' is shorter than the broken line PDP', by definition of a straight line. That is, according to the last paragraph, $2 \cdot \overline{PC}$ is shorter than $2 \cdot \overline{PD}$. Therefore PC is shorter than PD. 　—Q.E.D.

The length of the perpendicular from a point to a line is called the *distance* from the point to the line.

After drawing the perpendicular *PC* and any other line *PD* as in Fig. 24 it is apparent to the eye that *PC* is the shorter. But suppose the lines are extremely thin and *D* is extremely close to *C*; could it then be said at once that the perpendicular is the shorter? Unless there were some extremely accurate method of measuring the lines this question could not be answered directly. By letting *PD* represent *any* line other than the perpendicular so that *D* is *any* other point than *C*, however, we have proved logically and completely without any dependence on experimental trial that the perpendicular is shorter than *any* other such line, and therefore shorter than *all* such lines, that is, it is the *shortest*.

This example brings out the power and perfection of a geometrical proof.

This last theorem (XI) is related to X above and to III, article 24, and the three form an important group. Of the first two, the one states that of all the lines which can be drawn through a single point on a given line, only one is perpendicular to the line; and the other asserts the same of all the lines which can be drawn from an external point to the given line. Proposition XI then states that of all such lines drawn in either way, the perpendicular is the shortest. Perpendiculars are therefore unique among other lines.

30. Drawing of Perpendiculars. We next consider the problem of drawing perpendiculars. Practically, of course, perpendiculars (right angles) are drawn on a small scale by means of such devices as the carpenter's square or the T-square and triangle shown in Fig. 4, and on a very large scale by means of surveying instruments. But how were the carpenter's square, the T-square and the triangle themselves first formed? What are the geometric principles on which their construction is based? We shall see that the answer to these questions is contained in the principles stated and proved above.

In particular we get at once from IX the method for the solution of the following problem of construction:

XII. *To draw a perpendicular to a given line from a given external point.*

In Fig. 25 let *AB* be the given line and *P* the given external point, the perpendicular being not yet drawn.

Construction. Using the compass, with *P* as center and a radius sufficiently great, draw the circular arc CD cutting AB at C, D. Then, with center first at C and then at D, and with the same radius, greater than half of CD, draw the two arcs intersecting at E. Through P and this point E draw the line PE to meet AB. PE is the required perpendicular. —Q.E.F.

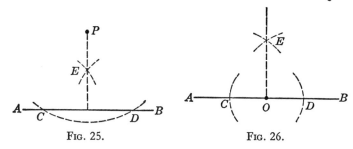

Fig. 25. Fig. 26.

Proof. Since by definition of a circle (article 14) all its radii are equal, the distance $PC = PD$ in Fig. 25, and also $CE = DE$. P is therefore equally distant from the ends of CD, and so also is E. According to IX therefore, P and E are on the perpendicular bisector of CD, and hence PE is perpendicular to AB. —Q.E.D.

NOTE. As in Figs. 24, 25 we shall use *solid* lines to represent given lines, and *dashed* lines for construction lines and required or resulting lines.

XIII. *To erect a perpendicular at a given point in a line.*

Construction. Let O be the given point in the line AB (Fig. 26). With O as center and any convenient radius describe the arcs intersecting AB at C, D. First with C and then with D as center, and with the same radius (greater than OC) describe arcs intersecting at E. Join O and E. OE is the required perpendicular. —Q.E.F.

Proof. Since O is the mid-point of CD, and since $CE = DE$, as radii of the same circle, then by IX, article 29, OE is the perpendicular bisector of CD. That is, $OE \perp AB$. —Q.E.D.

NOTE. If the given point O is at or near one end of AB a different method of solution is required. This depends on another property of the circle, which will be developed later.

XIV. *To bisect a given straight line.*

Let AB be the given line. From A and B as centers and with equal radii greater than half AB describe arcs intersecting at C and D. Draw CD intersecting AB at E. Then CD bisects AB at E.

For C and D are equidistant from the ends of AB by the construction, and hence, by IX, 29, CD is the perpendicular bisector of AB.

31. Parallel Lines. Straight lines which are in the same plane and do not meet however far they may be extended are called *parallel lines* and each is said to be parallel, or a parallel, to the others. Parallel lines are sometimes referred to simply as *parallels* The lines AB and CD, Fig. 28, are parallels. All the lines on an ordinary ruled sheet of writing paper are parallels. The symbol $\parallel$ is used for the words "parallel" or "is parallel to." Thus $AB \parallel CD$.

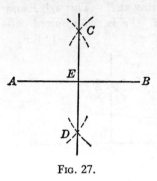

FIG. 27.

Euclid's famous *parallel postulate* (articles 9, 21) is the statement that *only one parallel to a given line can be drawn through a point outside the line.* This means that of all the lines drawn through a point, such as P, Fig. 29, one and only one can be parallel to the line AB; this is the line CD. All other lines through P will meet AB somewhere if extended far enough.

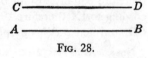

FIG. 28.

Parallel lines possess many interesting and valuable properties and bear certain relations to perpendicular lines. Thus,

XV. *Two straight lines in the same plane parallel to a third straight line are parallel to each other.*

For if they could meet, we should have two straight lines from the meeting point parallel to a straight line, but this is impossible according to the parallel postulate. Therefore they do not meet, and hence are parallel.

Similarly we can prove that

XVI. *Two straight lines in the same plane perpendicular to the same third line, are parallel.*

For if they are not parallel they would meet if extended far enough, and then we should have two lines from the same (meeting) point perpendicular to the same straight line, but this is impossible, according to X, 29.

Art. 31   PROPERTIES OF STRAIGHT LINES AND ANGLES   47

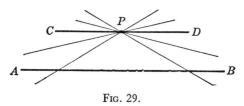

Fig. 29.

XVII. *If a straight line is perpendicular to one of two parallel lines it is perpendicular to the other also.*

In Fig. 30 let AB, CD be the two parallels, and let EF be perpendicular to AB at G and cut CD at H. We then have to prove that EF is perpendicular to CD.

Draw JK through H perpendicular to EF. (Until we know that CD is also perpendicular to EF we cannot say that JK will coincide with CD.) With JK thus drawn, then both AB and JK are perpendicular to EF, and hence, according to XVI, JK is parallel to AB. But CD, by hypothesis, is also parallel to AB, and as CD also passes through H, it must coincide with JK, since there cannot be two parallels to AB through H. But JK was drawn $\perp EF$, hence also $CD \perp EF$, or $EF \perp CD$, as was to be proved.

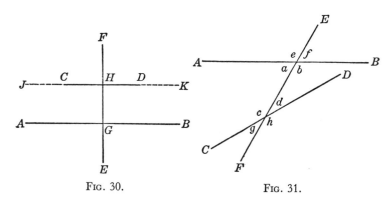

Fig. 30.   Fig. 31.

A straight line which cuts any two or more other straight lines is called a *transversal* of those lines. Thus in Fig. 31, and also in Fig. 30, EF is a transversal of AB and CD.

A transversal of two lines forms with these lines eight angles, as angles a, b, c, d, e, f, g, h in Fig. 31. The angles lying between the two lines which are cut by the transversal are called *interior* angles of the lines, and those lying outside are called *exterior* angles. Thus in Fig. 31 *a*, *b*, *c*, *d* are interior angles and *e*, *f*, *g*, *h* are exterior angles. These are related and distinguished in pairs as follows:

a and *d*, or *b* and *c*, are *alternate-interior* angles;
e and *h*, or *f* and *g*, are *alternate-exterior* angles;
e and *c*, *a* and *g*, *f* and *d*, or *b* and *h*, are *corresponding* angles.

It is seen at once that certain of the angles formed by the two lines and the transversal are equal. Thus the pairs of vertical angles, *a* and *f*, *e* and *b*, *g* and *d*, or *c* and *h*, are equal, as has already been shown (VII, 29).

If the two lines cut by the transversal are parallel there are several other pairs of equal angles, as we shall now see.

Thus in Fig. 32 let *AB*, *CD* be parallel lines cut by the transversal *EF* at *G* and *H*. Then $\angle a = \angle d$ and $\angle b = \angle c$. That is,

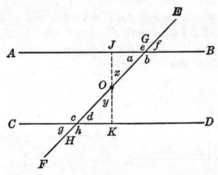

Fig. 32.

XVIII. *If two parallel lines are cut by a transversal the alternate-interior angles are equal.*

This is proved as follows: First draw *JK* through the middle point *O* of the segment *GH* and perpendicular to the two parallels (XVII), and forming the vertical $\angle x$, *y*. Then suppose the figure *HOK* placed on the figure *GOJ* with *OK* along *OJ*. Since the vertical $\angle x$, *y* are equal *OH* will then lie along *OG* (sides of equal angles coincide), and since by construction *OH* = *OG*, point *H* will lie on *G*.

Then the perpendicular *HK* will lie along the perpendicular *GJ*, because there can be only one perpendicular from *G* to *OJ*.

ART. 31 PROPERTIES OF STRAIGHT LINES AND ANGLES

Thus the two sides OH, HK of $\angle d$ coincide with the two sides OG, GJ of $\angle a$, and therefore by definition of equal angles, $\angle a = \angle d$, as was to be proved.

We can show in the same way that $\angle b = \angle c$, or more readily as follows, now that we know that $\angle a = \angle d$. Since all the lines are straight $\angle b$ is the supplement of $\angle a$ and $\angle c$ is the supplement of $\angle d$. But since $\angle a = \angle d$, b and c are both supplements of the same (equal) angles. Therefore $\angle b = \angle c$.

We have thus proved that the alternate-interior angles are equal. Conversely the following is also true and may be proved by a corresponding procedure:

XIX. *If two straight lines in the same plane are cut by a transversal and the alternate-interior angles are equal, the two lines are parallel.*

Consider now the *corresponding* angles d and f, Fig. 32. We have already shown that $\angle d = \angle a$; but as vertical angles, $\angle f = \angle a$. Angles d and f are therefore equal to the same angle, and hence, by the equality postulate, they are equal to each other: $\angle d = \angle f$. Similarly, it is easily seen that angles a, g are equal, as are also angles b, h and angles c, e. That is,

XX. *If two parallels are cut by a transversal, the corresponding angles are equal.*

From the same figure (Fig. 32) another relation among the angles is easily found. Thus, since angles b, f are supplementary and also angles d, f are equal, then b, d are supplementary; and in the same way, a, c are supplementary.

Similarly, it is easily shown that e, g and also f, h are supplementary. We can state therefore, as another parallel theorem:

XXI. *If two parallels are cut by a transversal, the two interior or exterior angles on the same side of the transversal are supplementary.*

The converse statements of XX, XXI are also true.

The theorems proved in this article show that parallel lines have as many interesting and unique properties as do perpendicular lines, and that parallels and perpendiculars are related. On examination it will appear that most of these unique properties of parallels and perpendiculars are results of the *parallel postulate* and the postulate stating that *all right angles are equal*.

In the next article some of these properties are used to demonstrate some further important properties of angles.

32. Some Further Properties of Angles. We begin with the theorem

XXII. *If the sides of two acute or two obtuse angles are parallel each to each, the angles are equal.*

In Fig. 33 let $\angle CAD$ and $\angle EBF$ be two acute angles with the sides $AC \parallel BE$ and $AD \parallel BF$, and call the angles respectively a and b.

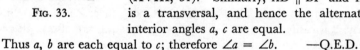

We then have to prove $\angle a = \angle b$.

First extend the sides AC, BF until they intersect, forming the acute angle c. Then by hypothesis $AC \parallel BE$, and BF is a transversal of the two parallels. The alternate-interior angles b, c are therefore equal (XVIII, 31). Similarly, $AD \parallel BF$ and AC is a transversal, and hence the alternate-interior angles a, c are equal.

Fig. 33.

Thus a, b are each equal to c; therefore $\angle a = \angle b$. —Q.E.D.

XXIII. *If the sides of two acute or two obtuse angles are perpendicular, each to each, the angles are equal.*

In Fig. 34 let $\angle CAD$ and $\angle EBF$ be two acute angles with sides $AC \perp BE$ and $AD \perp BF$, and call the angles respectively a and b.

We then have to prove $\angle a = \angle b$.

First draw $AG \parallel BE$ and $AH \parallel BF$, forming $\angle c$ with vertex at A. Since the sides of this angle are parallel to those of b, then angles b, c are equal, by the preceding theorem (XXII).

Now since $AC \perp BE$ and $AG \parallel BE$, then also $AG \perp AC$, by XVI, 31, and $\angle CAG$ is a right angle. Similarly $AH \perp AD$ and $\angle DAH$ is a right angle.

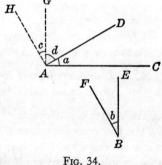

Fig. 34.

Next subtract the angle d from right angles DAH, CAG, leaving the equal remainders, angles a, c (since all right angles are equal). But

ART. 33 PROPERTIES OF STRAIGHT LINES AND ANGLES

we have already shown that angles b, c are equal. Therefore, $\angle a = \angle b$, as was to be proved.

If in Figs. 33, 34 the angles are both obtuse instead of acute, the same form of proof shows that they are equal in this case also.

If in Figs. 33, 34 the sides BF are drawn in the opposite direction from B, the sides of the angles will still be respectively parallel and perpendicular, but the angles at B will now be obtuse angles and equal to the supplements of the original angles b, and therefore also the supplements of the equal angles a. Therefore,

XXIV. *If the sides of an acute angle are parallel or perpendicular to those of an obtuse angle, each to each, the angles are supplementary.*

33. Drawing of Angles and Parallels. We take up now the problems of constructing angles and drawing parallel lines. The constructions are to be made without a T-square and without specifying the angle in degrees. Thus we have, first:

XXV. *At a given point in a given straight line, to construct an angle equal to a given angle.*

In Fig. 35 let DG be the given line and $\angle A$ the given angle. With D as vertex and DG as one side we then have to construct an angle equal to $\angle A$.

Construction. With A as center and any convenient radius AB describe an arc of a circle cutting the sides of $\angle A$ at B, C. With the same radius and with D as center draw an arc cutting DG at E.

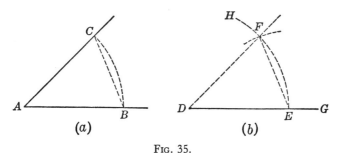

Fig. 35.

Next set the compass to the radius BC, Fig. 35(a), and with this radius and with E, Fig. 35(b), as center describe an arc cutting arc EH at F. Then draw DF.

$\angle EDF$ is the required angle, equal to $\angle BAC$. — Q.E.F.

This cannot be strictly proved until later, when we have first proved some propositions upon which this proof depends. It is easily seen to be true, however. Thus the arcs $\overset{\frown}{BC}$ and $\overset{\frown}{EF}$ are drawn with the same radius and so are parts of equal circles. Also, the equal straight lines BC, EF cut off equal parts of these circles, arcs $\overset{\frown}{BC}$ and $\overset{\frown}{EF}$. Therefore the angles A, D are equal parts of equal circles and hence, according to article 27, $\angle A = \angle D$.

XXVI. *To bisect a given angle.*

Construction. In Fig. 36 let $\angle AOB$ be the given angle, and with O as center and any convenient radius, draw arcs intersecting the sides of the angle at C, D and draw CD. Then take C and D as centers and with equal radii $CO = DO$ draw arcs intersecting at O and E and bisect CD at F, as in XIV, 30. Then OE bisects $\angle AOB$. —Q.E.F.

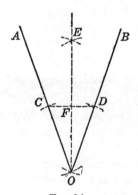

Fig. 36.

Proof. Since by the construction $FC = FD$ and O is a point on $OF \perp CD$, then according to VIII, 29, $\angle FOC = \angle FOD$, and hence OE divides $\angle AOB$ into two equal parts, that is, bisects $\angle AOB$. —Q.E.D.

By bisecting each of the angles FOC, FOD the original angle is divided into four equal parts; and by bisecting each of these, into eight equal parts, etc., but there is no general method of dividing an angle into any specified number of equal parts in general, or *multisecting* the angle, by ruler and compass. By means of special methods developed in analytical geometry |(article 8), however, any angle may be divided into any desired number of equal parts. These methods do not, however, make use of ruler and compass alone, as required in elementary Euclidean geometry.

In particular there is no ruler-and-compass method of *trisecting* an angle in general, although a few certain angles can be trisected by ruler and compass, and analytical geometry provides several methods for trisecting any angle.

Ruler-and-compass trisection is one of the three famous geometrical problems of antiquity (article 3) and it is only since the development of analytical geometry that it has been proved impossible.

Art. 33 PROPERTIES OF STRAIGHT LINES AND ANGLES

XXVII. *To draw a parallel to a given straight line through a given external point.*

In Fig. 37 let AB be the given line and P the given point outside AB. We then have to draw through P a straight line parallel to AB.

Construction. Through P draw an oblique line EF meeting AB at F. At P as vertex construct $\angle DPE = \angle BFP$, as in XXV, having CD as a side. Then $CD \parallel AB$.

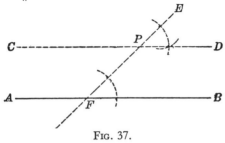

Fig. 37.

Proof. AB, CD are two straight lines cut by a line EF as a transversal, and by construction the corresponding angles BFP, DPE are equal. Hence $CD \parallel AB$. (XIX, 31.)

Parallel lines are drawn by draftsmen in two ways. One makes use of the construction given above, as shown in Fig. 38.

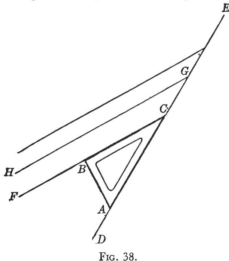

Fig. 38.

The draftsman's triangle ABC is laid on the paper with the longest side AC against the edge DE of the ruler or other triangle. By holding DE stationary and sliding the triangle along against it, the edge BC occupies successively the positions CF, GH, etc., and in any one of these positions a line is drawn along the edge BC of the triangle. These lines are parallel because $\angle HGC = \angle BCA$, the fixed angle of the triangle. These lines may be drawn in any desired position by adjusting the original position of DE.

If horizontal parallels on the drawing board are desired the T-square is simply placed in different positions on the board with the head held against the side of the board, and lines drawn along the edge of the blade as a ruler. Since the blade is perpendicular to the head and the side of the board, then all the lines are parallel, according to XVI, 31.

34. Size of the Earth. As an application of the principles XVIII, 31 and VI, 26 and the use of angle measure as defined in article 27 we will explain the method used by Eratosthenes in his measurement of the size of the earth (articles 3, 5).

Aristarchus (articles 3, 5) was the first of whom there is any record who taught that the earth is a round globe floating in space and not a flat disc floating on a vague unbounded sea, and Eratosthenes was the first to determine its size.

To begin with, the word "vertical" as used in common language with regard to direction on the earth, means "straight up-and-down" at any point on the earth's surface, that is, pointing towards the earth's center. Thus in Fig. 39, if the circle represents the round earth and each of the lines AD, BE, CF a stake or rod at the surface, then each of these is *vertical*. If the earth's surface were plane, line ABC would be straight and each of the lines AD, BE, etc., would be perpendicular to ABC and hence parallel to each other (XVI, 31). As it is, these lines are not parallel.

If the sun is vertically overhead at noon-time at any certain point on the earth's surface, then the rays shine vertically down and a vertical stake or pole at the place will cast no shadow. Now it is known that at the same noon instant the shadows of objects of the same length at different places have different lengths. Thus those in New York would be longer than those in Georgia, and would in the middle of the summer point due north. This is explained by reference to Fig. 40.

ART. 34   PROPERTIES OF STRAIGHT LINES AND ANGLES   55

The sun is so far away from the earth (nearly a hundred million miles) and is so much larger than the earth that the rays are in parallel lines and therefore the shadows are cast as in the figure.

Thus if the rays are vertical at A then the pole or object AD at A casts no shadow, that at B casts the shadow BG, and that at C the shadow CH.

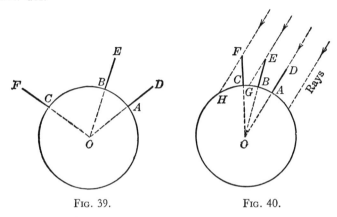

FIG. 39.   FIG. 40.

Eratosthenes utilized these known facts in his measurement. He found the place in Egypt (at Syene) where no shadows are cast at noon at the middle of the summer (the *solstice*) and precisely at noon of that day, the moment when the shadow is shortest, he measured the length of a vertical stake and also the length of its shadow, at Alexandria (articles 3, 4, 5), a known distance due north of Syene. The geometrical principles are shown in Fig. 41.

The distance AB (Syene to Alexandria) being known, and the height of the pole BE and length of the shadow BG being measured, the triangle GBE was then drawn and $\angle a$ measured. Now the ray DA is parallel to the ray EG. The pole-and-center line EBO then forms a transversal of these parallels, and according to XVIII, 31, $\angle b = \angle a$. Eratosthenes found this $\angle a$ to be very nearly 7 degrees 12 minutes. Therefore $\angle b = 7° 12'$.

Now $7° 12'$ is one 50th of $360°$. Therefore (article 27) the arc AB is $\frac{1}{50}$ the earth's circumference, and the distance around the earth is 50 times the distance AB. Expressed in modern units AB is about 500 miles. Eratosthenes thus found the circumference of the earth to

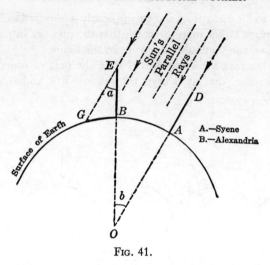

Fig. 41.

be about 25,000 miles. As the circumference of a circle (or sphere) is about $3\frac{1}{7}$ times the diameter, he found the diameter of the earth to be nearly 8000 miles.

35. Exercises.

1. Considering parallels as lines having the same direction, what is the angle between any two parallels?

2. State the reason why lines are parallel when drawn by a T-square which is placed in different positions by sliding the head along the straight edge of the board.

3. The sum of a pair of vertical angles is 70°. What is the magnitude in degrees of each of the pair of adjacent vertical angles?

4. Draw a series of several parallels as in Fig. 38, using the 30–60° draftsman's triangle with the 60° angle at A, and draw the line DE with the ruler in the position shown. (a) What is the value of each angle at C, G, etc.? (b) If a line is drawn perpendicular to DE and crossing the parallels, what angle will it make with each of them?

5. Refer to Fig. 32. If $\angle d$ is 40°, what is the value of each of the following angles: c, g, h, a, b, e, f, x, y?

6. Refer to Fig. 34. If $\angle a = 38°$, what are angles b, c, d?

7. Bisect an angle geometrically (Fig. 36). Measure the given angle and each part with the protractor and see how accurately the construction was executed.

8. By means of the protractor draw an angle of 64°. By repeated geometrical bisection divide the angle into eight equal parts, and check the accuracy of the construction by protractor.

ART. 35 PROPERTIES OF STRAIGHT LINES AND ANGLES

9. The circumference of the earth at the equator is very nearly exactly 24,912 land miles. The distance corresponding to one minute of angle on the equatorial circle is called a *nautical mile*. (a) How many nautical miles are there in the earth's equatorial circumference? (b) One nautical mile equals how many land miles? (c) How many land miles are equivalent to one degree of longitude at the equator? To one second?

10. The circumference of a circle is very nearly exactly 3.1416 times the diameter. What is the equatorial diameter of the earth?

Chapter 4

PLANE FIGURES FORMED BY STRAIGHT LINES AND ANGLES

36. Introduction. A geometric *figure* has been defined as a combination of points, lines and surfaces; and a *plane* figure as one lying entirely in one plane. We next consider a particular kind of plane figure, namely, *closed* figures.

Consider the four figures shown in Fig. 42. Here the figure (*a*) represents a simple acute angle; (*b*) shows three parallels cut by a

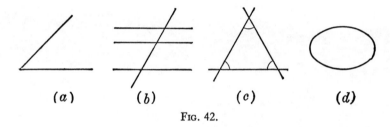

(*a*)   (*b*)   (*c*)   (*d*)

Fig. 42.

transversal; and (*c*), (*d*) show two forms of closed figures. The chief difference between these last figures and the first two is at once obvious: the closed figures surround or enclose a certain space in their plane, while this is not so with the others, which might in contradistinction be called *open* figures.

We have therefore as a concise definition: A *closed* plane figure is one which completely *encloses* a portion of its plane.

If a closed plane figure is formed of a curved line or lines, as in (*d*), it is sometimes called a *curvilinear* figure. Closed figures formed of straight lines, as in (*c*), are sometimes called *rectilinear* figures, and this term is also used to include such figures as (*a*) and (*b*).

In this chapter closed plane figures formed by straight lines and angles will be referred to simply as "figures."

The straight lines of which a closed rectilinear figure is formed are called its *sides*, and the angles formed by intersecting pairs of the sides

58

are called *the angles* of the figure; the vertex of one of these angles is also called a *vertex* of the figure. Thus in Fig. 42(c) the closed figure has three sides which form twelve angles. The three marked angles inside the figure are called *interior* angles when it is necessary to distinguish them from the others.

In Fig. 42(c) the length of each line segment of the sides included between the vertices of the angles is called the *length* of that side, and the sum of the lengths of the several sides of such a figure is called its *perimeter*. Closed rectilinear figures are usually drawn so that only these segments are given, as in Fig. 43. In such cases *the angles* of the

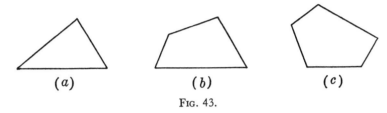

FIG. 43.

figure are taken as the interior angles. Thus in Fig. 43, (a) shows a figure of three sides and three angles, (b) a figure of four sides and four angles, etc. Such a figure as one of these has as many angles as it has sides.

Closed rectilinear plane figures are named according to the number of sides or angles they have, Latin or Greek root words being generally used. The Greek syllable *-gon* means "angle" and the Latin syllables *-lateral* mean *side*, and the first syllable or syllables of the name of a figure indicate the number of sides or angles.

Thus the three-sided or three-angled figure in Fig. 43(a) is called a *triangle;* that in (b) is a *quadrilateral;* that in (c) is a *pentagon;* etc. A general name for such a figure of any number of sides is *polygon* (*poly* = "several" or "many"), though this name is most often applied to figures of more than four sides.

In this chapter we take up the study of the properties of rectilinear closed plane figures under the separate headings *triangles, quadrilaterals* and *polygons*.

A. TRIANGLES

37. Forms of Triangles. Triangles may be of any shape, having sides of any length and having acute or obtuse angles.

If all the sides of a triangle have the same length, as in Fig. 44(a), it is called an *equilateral* triangle. In this case, as we shall see later, the three angles are also equal, and the triangle is called also *equiangular*. If two sides of a triangle are equal but the third side different it is called an *isosceles* triangle. Figure 44(b) represents an isosceles triangle. If no two sides of a triangle are equal it is called a *scalene* triangle. Figure 44(c) represents a scalene triangle.

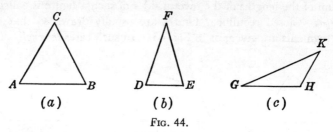

Fig. 44.

The vertex of an angle of a triangle is said to be a *vertex* of the triangle.

Triangles are usually marked and designated by capital letters placed at the vertices, as in Fig. 44, and the triangle is referred to by naming these letters, as "triangle ABC." Instead of the word "triangle" the symbol △ is much used; thus △ABC, and similarly in Fig. 44, △DEF, △GHK. The separate angles of a triangle are designated in the same manner as are any other angles.

Triangles are also classified according to their angles. If all the angles of a triangle are acute it is called an *acute* triangle, as in Fig. 45(a) below. A triangle which has an obtuse angle is called an *obtuse*

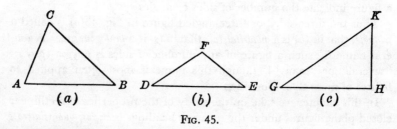

Fig. 45.

triangle, as in Fig. 45(b); and one which has a right angle is called a *right-angled* triangle, or simply a *right* triangle, as in (c).

In Fig. 45(a) each of the ∡A, B, C is acute; in Fig. 45(b) ∠F is the obtuse angle; and in (c) ∠H is the right angle.

ART. 38   PLANE FIGURES   61

The equilateral triangle (Fig. 44(a)) is also acute. The isosceles triangle may be acute or obtuse; that shown in Fig. 44(b) is acute. A scalene triangle may also be obtuse (Fig. 44(c)) or acute (Fig. 45(a)). An obtuse or a right triangle may be isosceles (when the two sides forming the obtuse or right angle are equal) but cannot be equilateral.

The two sides of a triangle whose intersection forms an angle of the triangle are said to *include* that angle. Each of these two sides is said to be *adjacent* to that angle; the third (other) side is said to be *opposite* the angle.

The two angles at the ends of a side are also sometimes said to include that side, and to be adjacent to it; and the other (third) angle is opposite that side.

By drawing the different forms of triangles described in this article, the reader may trace out many other relations among the forms of triangles. In a later article will be discussed problems of drawing triangles which must satisfy certain relations and conditions.

38. Triangle Descriptions and Definitions. A triangle may be drawn in any position whatever. If drawn with one side horizontal and below the rest of the figure, the triangle is said to *stand* on that side, and that side is called its *base*. Thus in Fig. 45(a), (b), (c) the triangles stand on the sides AB, DE, GH, respectively, as bases. Similarly in Fig. 46(a) AB is the base of △ABC. The point that is opposite the

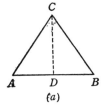 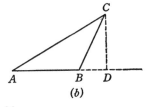

FIG. 46.

base is called the *vertex* of the triangle. In an isosceles triangle the vertex of the angle formed by the two equal sides is always called the *vertex angle* of the isosceles triangle. The two equal sides of the isosceles triangle are called its *legs*.

The line CD, Fig. 46(a), drawn from the vertex *perpendicular* to the base is called the *altitude* of the triangle. The *length* of this line is also sometimes called the altitude, or the *height*.

The altitude may be drawn inside the triangle, as in Fig. 46(a), or outside, as in (b), depending on the position in which the figure stands. Obviously a triangle may stand on any side as base. It may therefore have three altitudes. In any *one* position, without turning, all three of these lines may be drawn, each from a vertex perpendicular to the opposite side. Thus without regard to any base we have the general definition: An *altitude* of a triangle is a line drawn from a vertex perpendicular to the opposite side.

In addition to the altitudes, there are three other sets of lines connected with every triangle: the *medians;* the *bisectors of the angles;* and the *perpendicular bisectors of the sides.* The meaning of each of the last two terms is obvious. The definitions of all three are as follows:

A *median* of a triangle is a line drawn from a vertex to the *midpoint* of the opposite side. (A *median* and an *altitude* must not be confused.)

A line through a vertex of a triangle which divides the angle at that vertex into two equal parts is called an *angle bisector* of the triangle. (This is the usual meaning of the bisector of an angle.)

The *perpendicular bisector* of a side of a triangle is a line drawn perpendicular to the side at its midpoint. (This is the usual meaning of the perpendicular bisector of a line.)

These several important lines are shown in Fig. 47.

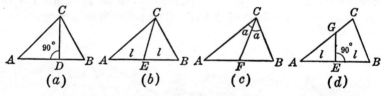

FIG. 47.

In Fig. 47(a) *CD* is the *altitude* on the side *AB*; in (b) *CE* is the *median* to side *AB*; in (c) *CF* is the *bisector* of ∠*C*; in (d) *EG* is the perpendicular *bisector* of side *AB*.

It is to be noted that in both (a) and (d) the angles at *D* and *E* are right angles (90°) but that in (d) the line through *E* does not pass through the opposite vertex *C*. Also in both (b) and (d) the segments *AE* = *BE* but again the line through *E* in (d) does not pass through *C*, while in (b) it does. Also in (b) it is the *side AB* which is bisected, while in (c) it is the opposite *angle C*; that is, in (b) the two angles at *C* are NOT equal, and in (c) the two parts of the side *AB* are NOT equal.

The distinctions between the altitude, median, angle bisector and perpendicular bisector of a side of a triangle can easily be made clear by drawing a fairly large scalene triangle and drawing the altitude, median, angle bisector, and perpendicular bisector of a side, all to the *same* side. No two of the four lines will be the same.

The equilateral (equiangular) triangle has the remarkable property, as we shall see later, that all four of these lines *are* one and the same on each of the three sides, and it is the only triangle which does possess this property.

The four sets of lines just described possess many remarkable properties, some of which we shall investigate in a later article.

Triangles in general possess many interesting and useful properties, some of which are taken up in the next article.

39. Some General Properties of Triangles. All plane triangles possess in common one remarkable property, regardless of size or shape. This property is, for *any* and *every* (Euclidean) plane triangle:

I. *The sum of the angles of a triangle equals a straight angle.*

This property of a triangle, or theorem, is also expressed by stating the sum to be equal to *two right angles* or to 180 *degrees*.

It is easily proved, as follows:

In Fig. 48 let ABC be any triangle, with $\angle A$, B, C. We then have to prove that $\angle A + \angle B + \angle C = 1$ st. $\angle = 180°$.

Draw the line DE through the vertex C parallel to the opposite side AB; then AC and BC are transversals cutting two parallels.

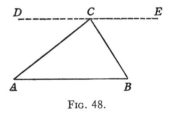

Fig. 48.

The alternate-interior angles A and DCA are then equal, and also the alternate-interior angles B and ECB (by XVIII, 31). Therefore the three angles DCA, ACB, BCE at C are the same as the three angles of the original triangle. But the sum of the three adjacent angles at C is the straight angle DCE. Therefore the sum of the angles of the triangle equals one straight angle, or $\angle A + \angle B + \angle C = 180°$, as was to be proved.

This is one of the most important properties of a triangle and is one of the foundations of the subject of trigonometry.

From this property we can at once obtain several other important properties of any triangle. Thus since an obtuse angle is one which is greater than 90° a triangle can obviously have but one obtuse angle, for if one angle is greater than 90° the sum of the other two must be less than 90° and therefore both must be acute. Any triangle may, however, have three acute angles.

If the three angles of a triangle are equal, then each is one-third of the total, that is, $180 \div 3 = 60°$. Therefore,

   II. *Each angle of an equiangular (equilateral) triangle is 60°.*

If a side of a triangle be extended beyond a vertex, as BD, Fig. 49, the new angle formed at that vertex, as ∠CBD, is called an *exterior* angle of the triangle. If all three of the sides are thus extended *in the same order*, going around the triangle, the three exterior angles so formed are called *the* exterior angles of the triangle.

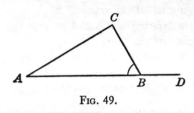

Fig. 49.

The two angles A, C, Fig. 49, of the triangle are called the *opposite interior* angles to the exterior ∠CBD.

Now the interior ∠B + ∠CBD equals the straight ∠ABD. But ∠B + (∠A + ∠C) = 1 st. ∠ also, as just proved above (I). Therefore, ∠CBD = ∠A + ∠C. That is,

   III. *An exterior angle of a triangle equals the sum of the two opposite interior angles.*

The sides and angles of a triangle are called its *parts*.

If the three sides and three angles of one triangle are exactly equal to the three sides and the three angles of another triangle, so that either can be laid exactly on the other and the two coincide, part for part, the two triangles are said to be *congruent triangles*, and each is *congruent to* the other.

In general, it is not necessary to test all the six parts of each of two triangles to find whether they are congruent or not. If it is known that three parts of one (one of which must be a side) are the same as three corresponding parts of the other, then it can be shown to be a necessary consequence that the other three parts are also equal in the two, and therefore the two triangles are congruent. Thus

   IV. *If two sides and the included angle of one triangle are equal to two sides and the included angle of another triangle, the triangles are congruent.*

In Fig. 50 let (*a*) and (*b*) represent the two triangles, with sides AB, AC equal to DE, DF respectively and $\angle A = \angle D$. In order to show that the triangles are congruent we then have to prove that $\angle B = \angle E$, $\angle C = \angle F$ and $BC = EF$.

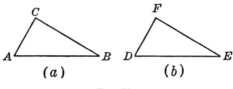

Fig. 50.

In order to prove these relations, first suppose $\triangle ABC$ placed on $\triangle DEF$ so that the vertices A, D and sides of the equal angles coincide (as they will, by definition of equal angles). Then, since by hypothesis $AB = DE$ and $AC = DF$, B will lie on E and C on F. Then since there can be only one straight line joining two points, BC must coincide with EF. That is, $BC = EF$, and since their vertices and sides now coincide, $\angle B = \angle E$ and $\angle C = \angle F$, as was to be proved.

Similarly, we can prove that

V. *If two angles and the included side of one triangle are equal to two angles and the included side of another, the triangles are congruent.*

Thus in Fig. 50 let (*a*) and (*b*) represent the two triangles with now $\angle A = \angle D$, $\angle B = \angle E$ and the sides $AB = DE$. In order to show the triangles are congruent we now have to prove that the sides and vertices of $\angle C$, F coincide, $AC = DF$ and $BC = EF$.

Suppose $\triangle ABC$ placed on $\triangle DEF$ so that AB coincides with its equal DE. Then since $\angle A = \angle D$ and $\angle B = \angle E$, by hypothesis, AC lies along DF and BC along EF. Since two straight lines can intersect in only one point, then C will lie on F. Hence the triangles will coincide in every part and are congruent, as was to be proved.

By proceeding in a manner similar to that used above it is also readily proved that

VI. *If a side and any two angles of one triangle are equal to the corresponding side and two angles of another, the two triangles are congruent.*

An *isosceles* triangle has been defined as one having two equal sides, called its *legs*. It can now be proved that

VII. *In an isosceles triangle the angles opposite the legs are equal.*

Let the triangle of Fig. 51 represent any isosceles triangle, with vertex A and legs $AB = AC$. We have to prove that $\angle B = \angle C$.

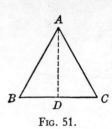

Fig. 51.

Draw AD so as to bisect $\angle A$, forming two triangles ADB, ADC.

Then in the two triangles, $\angle BAD = \angle CAD$, by definition of bisection; sides $AB = AC$, by hypothesis; and also $AD = AD$, as the same identical line in both triangles.

That is, two sides and the included angle of $\triangle ADC$ equal two sides and the included angle of $\triangle ADB$, and hence, by IV, the two triangles are congruent. Therefore, as corresponding parts of congruent triangles, $\angle B = \angle C$, as was to be proved.

Since any two sides of an equilateral triangle are equal, therefore any two angles (opposite these sides) are equal. Therefore all three of the angles are equal and

VIII. *An equilateral triangle is also equiangular.*

In a manner similar to that in which VII was proved it can be shown also, as would be expected, that

IX. *If two angles of a triangle are equal, their opposite sides are equal, and the triangle is isosceles.*

From the results already found, the following interesting property of the isosceles triangle can now be demonstrated:

X. *The altitude from the vertex of an isosceles triangle is also the median, the bisector of the vertex angle, and the perpendicular bisector of the base.*

In Fig. 51 AD is drawn to bisect $\angle A$, and it was shown that the two triangles ADB, ADC are congruent. Therefore the corresponding angles BDA, CDA are equal, and hence by definition, $AD \perp BC$; that is, AD is the altitude on BC.

Also, in the congruent triangles ADB, ADC, side $BD = CD$; that is, AD bisects BC and is therefore the median from A.

Since $AD \perp BC$ and AD bisects BC, it is therefore also the perpendicular bisector of BC, and the entire proposition is proved.

There is one more important condition which determines the congruence of triangles. It is:

XI. *If the three sides of one triangle are equal respectively to the three sides of another, the triangles are congruent.*

This statement should be true almost obviously, but it might be conceivable that three given lines of specified lengths could be put together in different ways to form different triangles. It is therefore necessary to prove that only one such triangle is possible, or, what amounts to the same thing, if two triangles are so formed from the same sides, they will be congruent.

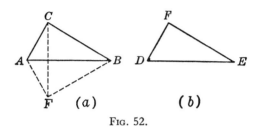

FIG. 52.

To prove this, in Fig. 52 let ABC, DEF be two triangles with sides $AB = DE$, $BC = EF$, $CA = FD$. We then have to show that the two triangles are congruent.

Suppose $\triangle DEF$ placed against $\triangle ABC$, as in Fig. 52(a), so that DE coincides with its equal AB and vertex F is opposite C; and draw CF.

Now since in (a) $AC = AF$ by hypothesis, $\triangle CAF$ is isosceles and hence $\angle ACF = \angle AFC$, by VII above. Similarly $BC = BF$, $\triangle CBF$ is isosceles, and $\angle BCF = \angle BFC$.

Therefore, by the equality addition postulate,

$$\angle ACF + \angle BCF = \angle AFC + \angle BFC;$$

that is,

$$\angle ACB = \angle AFB.$$

Therefore the $\angle ACB$ and its including sides are equal to $\angle AFB$ ($\angle DFE$) and its including sides. According to IV above, therefore, the two triangles are congruent, as was to be proved.

The properties expressed in theorems IV, V, VI and XI are very important. They state in effect that if any three parts (except the three angles alone) of a triangle are given or known the triangle is completely determined. This means that with three such parts given the triangle can be constructed geometrically, as in article 42 below, and the remaining three parts can also be calculated without drawing, by the methods of trigonometry.

That the angles alone are not sufficient to determine the complete triangle is obvious when it is remembered that the angles simply determine the *shape* of the figure and not its *size*, as the angles have nothing to do with the lengths of the sides when the three angles are all known.

The *bisector* of an angle and the *distance* from a point to a line (the perpendicular) have already been defined. By the aid of some of the preceding theorems the following interesting property of an angle bisector can now be demonstrated:

XII. *Any point in the bisector of an angle is equally distant from the two sides of the angle.*

In Fig. 53 let O be any point in the bisector AD of $\angle BAC$. OE and OF are then the distances from O to the sides of the angle, and we have to prove that $OE = OF$.

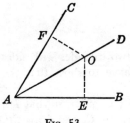

Fig. 53.

Since by hypothesis $\angle OAE = \angle OAF$ and by definition of distance to a line, $\angle OEA = \angle OFA =$ rt. $\angle$, the remaining angles in the triangles AOE, AOF must be equal (since their sum in each case is 180°). That is, $\angle OAE$, AOE are equal respectively to $\angle OAF$, AOF. Also in the two triangles, side $AO = AO$ as an identity. Therefore the two triangles have two angles and the included side respectively equal, and hence are congruent (by V). The corresponding sides OE, OF are therefore equal, as was to be proved.

This theorem should be compared with IX, 29. These two are companion theorems.

Triangles are closely related to quadrilaterals and several additional general properties of triangles will be developed later, in the section (B) on quadrilaterals. In the next article we take up a few interesting special properties of triangles.

40. Some Properties of Altitudes, Medians, Angle Bisectors and Perpendicular Bisectors of Sides of Triangles. The special lines associated with triangles, discussed in article 38, have some remarkable properties. The first of these is

ART. 40   PLANE FIGURES   69

XIII. *The three bisectors of the angles of a triangle meet in one point, which is equidistant from the three sides.*

In Fig. 54 let ABC represent any triangle and let AD, BE, CF respectively be the three bisectors of its angles A, B, C. We then have to prove that AD, BE, CF meet in one point, and that this point is equally distant from the sides AB, BC and CA.

First consider only AD and BE. Since these are not parallel, they intersect in some point, say at O. Then, since O is on AD the bisector of $\angle A$ it is equidistant from AB and AC, according to XII, 39. Similarly O is on BE and hence equidistant from AB and BC. O being thus equidistant from AC and BC, it therefore lies on CF the bisector of $\angle C$, which thus passes through the intersection of AD and BE. Thus the three angle bisectors meet in the single point O which is equidistant from the three sides.

XIV. *The three perpendicular bisectors of the sides of a triangle meet in one point which is equidistant from the three vertices.*

In Fig. 55 let DG, EH, FK be perpendicular to the sides AB, BC, CA, respectively, of $\triangle ABC$, at their midpoints D, E, F. To prove: that

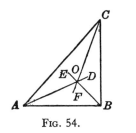
FIG. 54.

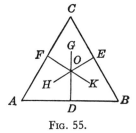
FIG. 55.

DG, EH, FK meet in one point O which is equally distant from A, B and C.

First consider only DG and EH. Since these are not parallel they intersect, say at O. Now O is on DG the perpendicular bisector of AB. According to IX, 29, therefore, O is equally distant from A and B. Similarly O is on EH the perpendicular bisector of BC and hence is equidistant from B and C. Thus O is equally distant from A and C, and is therefore on FK, the perpendicular bisector of CA, which thus passes through the intersection of DG and EH. Thus the three perpendicular bisectors of the sides meet in the single point O, which is equidistant from the three vertices.

Two other remarkable properties are expressed in the following theorems:

> XV. *The three altitudes of a triangle meet in one point.*
>
> XVI. *The three medians of a triangle meet in one point whose distance from each vertex is two-thirds the median from that vertex.*

The proofs of these two theorems depend on certain properties of quadrilaterals, and will be given in section B.

The point of intersection of the medians of a triangle is called the *center of gravity* of the triangle. If a triangle is cut from paper, board, etc., of uniform thickness and weight, it can be balanced on a single supporting point placed at the center of gravity.

If three or more lines pass through one point they are said to be *concurrent*. The three lines of each group associated with a triangle in the last four theorems therefore concur.

Since the equilateral (equiangular) triangle is symmetrical in every respect, the four theorems above all apply to it at the same time and can be combined into one statement, which is an extension of X, 39. We have, namely, that

> XVII. *The altitudes, medians, angle bisectors, and perpendicular bisectors of the sides of an equilateral (equiangular) triangle all coincide and meet at a point two-thirds the distance from each vertex to the opposite side.*

41. Right Triangles. A *right triangle* has been defined as one which has a right angle.

Right triangles have a number of interesting and important properties not possessed by other forms of triangles. A number of these properties have to do with the three angles and their relations.

Since the sum of the angles of any triangle is two right angles, a triangle can have only *one* right angle. The sum of the other *two* angles is therefore one right angle. Therefore

> *A right triangle has one right angle and two acute angles. The two acute angles of a right triangle are complementary.*

These statements are illustrated in Fig. 56.

Here the right angle in (*a*) is C, $= 90°$. Therefore $A + B = 90°$, and $A = 90° - B$ or $B = 90° - A$. If, for example, $\angle A$ is $30°$ then

∠B is 60°. Similarly in Fig. 56(b), where the right triangle is isosceles, with A, B the equal angles opposite the equal sides, A and B are each 45°.

The 30–60° and the 45° right triangles are cut from celluloid or similar material or made of wood for the use of draftsmen in drawing parallels and in many other geometrical constructions (see Figs. 4, 38).

By drawing with these triangles and the T-square, angles which are the sums and differences of 30, 45, 60 and 90 degrees, acute and obtuse angles up to 135°, varying in steps of 15°, are easily constructed. Thus the 30° angle bisected (XXV, 33) gives the 15° angle. Also by drawing the 45° angle and subtracting 30° from it by means of the 30° triangle, the 15° angle is obtained. The triangles furnish the 30, 45, 60, 90° angles directly and the sum of the 30° and 45° gives the 75° angle.

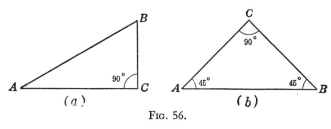

FIG. 56.

The two perpendicular sides of a right triangle are called the *legs* of the right triangle. They are the sides opposite the acute angles. The third side, opposite the right angle, is called the *hypotenuse*. (This term must not be confused with the *hypothesis* of a theorem.) Thus in Fig. 56 AC and BC are the legs of the right $\triangle ABC$ and AB is the hypotenuse.

If a right triangle stands on either leg as base, the other leg, being perpendicular to it, is the altitude on that side. Thus in Fig. 56(a) the leg AC is the base and the leg BC is the altitude on that side. In the isosceles right triangle of Fig. 56(b) the hypotenuse AB is the base and the corresponding altitude is the perpendicular bisector of AB (X, 39).

We state here, without proof, the following property of the 30–60° right triangle; it will be proved later.

The hypotenuse of the 30–60° *right triangle is twice the shorter leg.*

From the general theorems on the congruence of triangles in article 39 may be obtained at once the several corresponding theorems for right triangles. Thus, since the two legs include the right angle, and in any right triangle the right angles are of course equal, theorem IV, 39, becomes, for the right triangle:

XVIII. *If the legs of one right triangle are equal to those of another right triangle, the triangles are congruent.*

Since in any two right triangles the right angles are always equal, theorems V, VI, 39, give also:

XIX. *Two right triangles are congruent if a leg and an acute angle of one are equal to a leg and the corresponding acute angle of the other.*

XX. *Two right triangles are congruent if the hypotenuse and an acute angle of one are equal to the hypotenuse and an acute angle of the other.*

It is easy to prove also the following:

XXI. *If a leg and the hypotenuse of one right triangle are equal to a leg and the hypotenuse of another right triangle, the two triangles are congruent.*

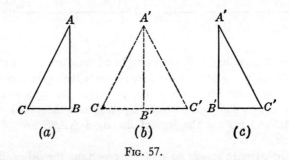

FIG. 57.

Thus in the right triangles ABC, $A'B'C'$, Fig. 57(a), (c), let the hypotenuse $AC = A'C'$ and the leg $AB = A'B'$. To prove: that triangles ABC, $A'B'C'$ are congruent.

Suppose $\triangle ABC$ placed against $\triangle A'B'C'$ as in Fig. 57(b), so that the equal legs AB and $A'B'$ coincide, A falling on A' and B on B'. Then since the original $\angle B$, B' are right angles, $A'B' \perp CC'$ and CC' is a straight line, so that $A'CC'$ is a complete triangle.

Also $\triangle A'CC'$ is isosceles, because by hypothesis the hypotenuses AC, $A'C'$ are equal. Therefore $\angle C = \angle C'$, according to VII, 39.

We thus have the hypotenuse AC and acute angle C of $\triangle ABC$ equal to the hypotenuse $A'C'$ and acute angle C' of $\triangle A'B'C'$, and hence the triangles are congruent (XX).

The properties of right triangles set forth in the preceding propositions are of the highest importance in geometry and trigonometry. Several other interesting and important properties will be demonstrated later, when we have proved some preliminary theorems on which they depend. One of these is the remarkable property expressed by the famous *Theorem of Pythagoras* (article 3), which we state here:

XXII. *The square of the hypotenuse of a right triangle equals the sum of the squares of the legs.*

In the case of such a right triangle as those of Fig. 56 this property can be expressed algebraically as

$$\overline{AB}^2 = \overline{AC}^2 + \overline{BC}^2$$

where $\overline{AB}$, etc., indicate the measured *lengths* of the sides AB, etc.

42. Triangle Constructions. In the following problems no lines or angles are specified by stating the length in inches, etc., or the angle in degrees, but the line or angle itself is simply given already drawn. It is to be used or duplicated in the required construction.

We can now complete the proof of the construction of an angle, given in XXV, 33. Referring to Fig. 35, constructed as there described, DE and DF are equal to AB and AC, as they are marked off with the same compass setting; and $EF = BC$, for the same reason. The three sides of $\triangle DEF$ are thus equal to the three sides of $\triangle ABC$, and hence the triangles are congruent, according to XI, 39. The corresponding $\angle A$, D of the triangles are therefore equal, as was to be proved.

By similar use of the triangle definitions and theorems of this chapter the following construction problem may be solved. Thus let it be required

XXIII. *To find the third angle of a triangle when the other two are given.*

If the two angles are given by stating their values in degrees it is only necessary to add them and subtract the sum from 180°, since the sum of the three angles of a triangle is 180°, a straight angle (I,39).

In the present problem, however, the actual angles are given, already drawn, as in Fig. 58(*a*), (*b*). Let these be $\angle A = a$ and $\angle B = b$ of a triangle ABC. The problem is then to find by *geometrical construction* the third angle $\angle C = c$.

Using any convenient radius and with A, B as centers draw the arcs as in Fig. 58(a), (b) and with the same radius and any point C on any line DE as center, draw an arc intersecting DE, as in Fig. 58(c). Continuing as in XXV, 33, construct $\angle ECF = a$ with C as vertex and CE as a side, and similarly with C as vertex and CF as a side construct $\angle FCG = b$. $\angle DCG = c$ is then the required third angle C of the triangle.

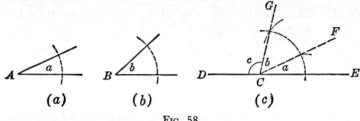

Fig. 58.

For, in Fig. 58(c), $a + b + c = 1$ st. $\angle = \angle A + \angle B + \angle C$, and by construction $a = \angle A$, $b = \angle B$; hence $c = \angle C$.

XXIV. *To construct a triangle when two sides and their included angle are given.*

In Fig. 59(a) let b, c be the given sides and $\angle A$ the given included angle.

Lay off $AB = c$, Fig. 59(b), and at A construct $\angle BAD = \angle A$. On AD lay off $AC = b$, and draw BC. $\triangle ABC$ is the required triangle, for it has the two given sides and included angle.

In Fig. 59(b) the third side BC may be marked as a. Here BC, CA, AB indicate the *lines* which form the sides of the triangle and a, b, c are

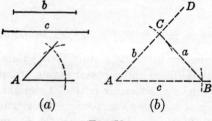

Fig. 59.

algebraic symbols which stand for the numbers which represent the *lengths* of the sides. This scheme is generally used, with side *a* opposite angle *A*, side *b* opposite *B*, and side *c* opposite *C*.

XXV. *To construct a triangle when a side and two angles of the triangle are given.*

In Fig. 60(*a*) let ∡*A*, *B* be the given angles and *c* the given side. Lay off the line $AB = c$, as in Fig. 60(*b*). At *A* construct $\angle BAF = \angle A$ and at *B* construct $\angle ABG = \angle B$. Extend *AF* and *BG* until they meet at *C*. Then △*ABC* is the required triangle.

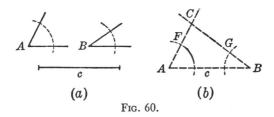

(*a*)   (*b*)

Fig. 60.

For it has the two given angles and included given side, and according to V, 39, is the same as the specified triangle.

XXVI. *To construct a triangle when the three sides are given.*

In Fig. 61(*a*) let *a*, *b*, *c* be the given sides.

As in Fig. 61(*b*) lay off $AB = c$. With *B* and *A* as centers and radii respectively equal to *a* and *b*, draw arcs intersecting at *C*. Draw *AC* and *BC*. Then △*ABC* is the required triangle, for it has the three given sides.

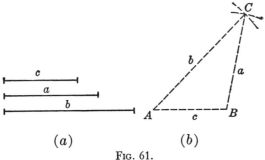

(*a*)   (*b*)

Fig. 61.

There is one other general case in which a triangle may be constructed when three parts are given. This is

XXVII. *To construct a triangle when two sides and the angle opposite one of them are given.*

The complete solution of this problem is somewhat involved and will not be given here. Depending on the relative lengths of the sides there are three cases, according as $a = b$, $a > b$, $a < b$, where a, b are the two given sides. In each of the first two of these three cases there is only one solution. In the third case there are two solutions if the given angle is acute, and no solution if it is obtuse.

Readers interested in the full solution and discussion of this problem will find it treated in the book of Wentworth mentioned in article 7 and that of Phillips and Fisher mentioned in article 9.

The construction of right triangles and isosceles triangles involves no new methods not already given. Several of the steps in the constructions given above are modified and generally simplified, however, on account of the facts that two of the sides of the right triangle are perpendicular and two of the sides of the isosceles triangle are equal with equal opposite angles.

These constructions are left as exercises for the reader. In these the ruler and compass alone should be used (articles 13 and 14).

43. Illustrations and Applications. The arts and trades of drawing, carpentry, architecture, machine work, building, surveying, etc., might almost be taken entirely as illustrations of the principles and methods of geometry. For these, however, reference must in general be made to special books on those subjects or their different branches.

In this book we shall give only such illustrations and applications from the common arts or trades and from everyday observation as may be understood by one who has no special technical knowledge, and which will serve clearly to bring out the meaning of the geometrical principle or method involved.

Thus, we have already described in the appropriate places in the preceding chapter the applications of certain properties of parallels and perpendiculars in drawing; and Eratosthenes' measurement of the earth illustrates how an extremely important result may be obtained by the use of a very simple principle.

In this article we shall present illustrations and applications which involve properties of triangles only.

Example 1. In laying out a garden, or a tennis court or the lines of the walls of a building, or the like, it is necessary to run straight lines of considerable length. Here the methods of the drawing board will not serve and such lines are usually laid out by stretching a light string from one to the other of two points on the required line. Here the stretched string is the *shortest line joining the two points* and is therefore *straight*. This method is generally used by gardeners, surveyors, and builders.

Example 2. In drawing a perpendicular to a straight line run as in Example 1 the method of XII, 30, is used, a stretched string of the proper length being swung about the fixed centers to draw the arcs, instead of using the compass.

Example 3. To run a parallel to a straight line laid out as in Example 1 the method illustrated in Fig. 62 is very convenient.

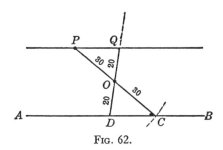

Fig. 62.

Suppose the line AB is already run, and it is required to run a line through a specified point P parallel to AB. Stretch a long string or measuring tape, say 60 feet, from P and swing it around P until the other end meets AB at C. Mark the middle of PC at O and from any convenient point D on AB run another line through O. Measure DO; suppose it is 20 feet. Then along the same line lay off from O the same length (20) to a point Q. Run a straight line through P and Q. PQ is the desired parallel to AB.

For, the vertical $\angle POQ = \angle COD$ at O and in the two triangles OPQ, OCD the sides OP, OQ are equal to the sides OC, OD respectively. Hence the triangles are congruent (IV, 39), and therefore by definition of congruent triangles, the corresponding angles PQO, CDO are equal. Hence $PQ \parallel AB$ (XIX, 31).

Example 4. Cut out a triangle nearly as large as a sheet of paper and mark its angles A, B, C as in Fig. 63(a).

Tear off the corners and fit them together as in (b) with the vertices together at one point. A straight line is then formed along the bottom edge of the whole.

For, the sum of the three angles is 180°, a straight angle (I, 39), and the external sides must lie in a straight angle (straight line) when placed adjacent (article 25).

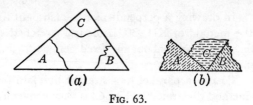

FIG. 63.

Example 5. In Fig. 64 AB represents an irregular object or area such as a pond, large pile of stones, a building, etc., such that the line AB cannot be measured straight through. How can the distance AB be determined by a simple measurement?

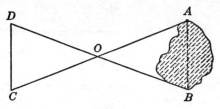

FIG. 64.

Drive a stake at some convenient point O, measure the distances AO and BO, sight along AO and BO and set stakes at C and D so that $OC = OA$ and $OD = OB$. Then $CD = AB$ and is easily measured.

For, vertical $\angle AOB = \angle COD$ and as the sides OC, OD equal OA, OB respectively, the two triangles are congruent (IV, 39), and hence the corresponding parts $CD = AB$.

Example 6. Suppose it is desired to find the distance of an island or a raft R from a point P on the shore of a body of water (Fig. 65) when R is inaccessible. A simple method is given below.

From P lay off the line PQ of a convenient length and so that R can be seen from Q. From both P and Q sight on the same point R and mark the lines of sight PR, RQ, forming the angles a and b with PQ.

Then form the same angles on the other side of PQ as shown and extend the sides of these angles until they meet at S. Then $PS = PR$ and is easily measured.

This is at once seen to be true because the angles a, b, and included side PQ in $\triangle PQS$ are the same as in $\triangle PQR$, and hence the triangles are congruent (V, 39). Therefore the corresponding sides $PS = PR$.

This principle is very easily applied and has been in use since very early times. Thales (article 3) is said to have used it to determine the distance from shore of the anchored ships of the enemies of the Greeks in one of their early wars.

Example 7. The principle of Example 6 is used in a slightly different manner in Fig. 66 to measure the distance to another inaccessible point, say across a river.

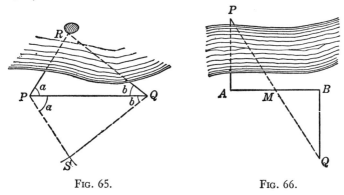

Fig. 65.   Fig. 66.

From the point A directly opposite P walk directly to B and count the steps, AB making a right angle (easily estimated) with AP. Set a stake at the middle point M of AB. Walk directly away from B (at a right angle) and count the steps until a point Q is reached which is directly in line with M and P. Then $BQ = AP$.

For, MAP and MBQ are triangles with two angles and included side respectively equal ($\angle A = \angle B$, $\angle AMP = \angle BMQ$, $AM = BM$), or they are right triangles with one leg and acute angle equal ($AM = BM$, $\angle AMP = \angle BMQ$). Hence as in Example 6 or by XIX, 41 the triangles are congruent and the corresponding parts BQ, AP are equal.

Example 8. The principle used in Example 6 is very easily adapted to a variety of measurements of inaccessible distances and has since the earliest times been a favorite with surveyors and military engineers and observers. A method which is fairly accurate is shown in Fig. 67(a) and a less accurate but very rapid adaptation in Fig. 67(b).

In Fig. 67(a) AC is a vertical rod or staff with another (cross) staff, having a straight edge, fastened to it by a single screw or nail. The cross piece is adjusted and sighted so that it points to the inaccessible point B. The rod is then turned while remaining vertical, without changing the setting of the cross piece, and a sight is taken along the inclined cross piece, toward D where the line of sight strikes the level ground. The two marked angles at A and C and the included height AC of the rod remaining the same, AD is then equal to AB.

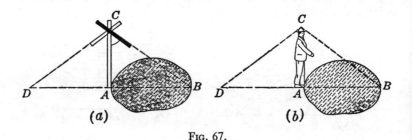

Fig. 67.

In Fig. 67(b) a man stands at A and sights past the edge of his hat brim or cap visor at B, and without raising or lowering his head turns and again sights past the edge toward D where the line of sight meets the level ground. As in (a), $AD = AB$.

It is said that by a quick use of this method a young military engineer in Napoleon's army determined the width of a stream which blocked a march, and so won the Emperor's high favor for his quick thinking and originality.

Example 9. In cutting the rafters for a gable roof, such as that illustrated in Fig. 68, a carpenter, whether he does so consciously or not, actually makes use of a property of the isosceles triangle.

The rafters are of the same length and hence $\triangle ABC$ is isosceles. Therefore $\angle B = \angle C$ (VII, 39), and hence the angle b at which one rafter is sawed to fit the corner or edge of the beam BC, is the same as $\angle c$.

Also, when it is desired to run a tie rod from A to D to support a part of the load on the beam, if the rod is vertical it will be fastened at the center of BC (X, 39) and so bear directly upward and equally divide the load on the beam.

In roof construction such as shown in Fig. 68 the base BC of the isosceles triangle is called the *span* of the roof; half the span (CD) is called the *run* of the rafter; and the altitude AD is called the *rise*. The relation of the rise to the run determines the steepness or flatness of the roof and the quotient of the rise divided by the span is called the *pitch* of the roof. Thus *pitch* = (*rise*) ÷ (*span*). For some purposes the pitch is defined by the relation: *pitch* = (*rise*) ÷ (*run*).

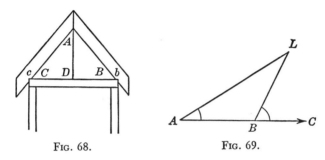

FIG. 68.   FIG. 69.

Example 10. Navigators use a method for determining distances at sea without stopping the ship, which utilizes a property of the isosceles triangle together with the principle III, 39. Thus in Fig. 69 suppose the ship is following the straight course ABC and L is a point, such as a lighthouse, whose distance from the ship is required.

When the ship is at the position A the angle CAL between the course AC and the line of sight AL to L is measured. The distance travelled by the ship is then measured from A to a point B where $\angle CBL$ is just twice $\angle CAL$. Then the distance $BL = AB$.

For, according to III, 39, the exterior angle CBL equals the sum $\angle A + \angle L$. Hence when $\angle CBL = 2\angle A$, then $\angle A = \angle L$ and by IX, 39, $BL = AB$.

This method of measuring distances is known as "doubling the angle on the bow," the angle between the line of the course, straight ahead over the bow of the ship, and the line of sight to a point on either side, being called the "angle on the bow."

Example 11. Another property of the isosceles triangle is very easily utilized to do very accurate levelling without a spirit level. This depends also on the fact in nature that gravity acts, at any point on the earth, in a straight line toward the center of the earth. This line is called the *vertical* at that place, and a line perpendicular to it is called the *horizontal* or *level* at that place (see articles 34 and 24). If a weight is suspended at one end of a string which is supported at the other end, the string will be in the vertical. This device is called a *plumb bob*. Any instrument for determining the level is called a *level*.

A very simple and accurate level is easily made as in Fig. 70. An isosceles triangle of convenient size is made of two pieces of wood as the legs (or cut from cardboard) with a cross piece forming a letter A.

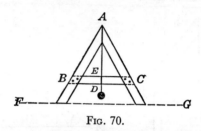

Fig. 70.

This piece must be parallel to the base of the large triangle, forming also an isosceles triangle ABC. Across the exact center of the bar BC the line DE is drawn, and a plumb bob is suspended from the apex A. When the frame is set up on a surface or line FG so that the plumb line hangs freely on DE, then FG is level, or horizontal.

For, AD is vertical and D is the middle of BC, and hence in isosceles $\triangle ABC$, $BC \perp AD$ (X, 39). Since $FG \parallel BC$ by construction, then also $FG \perp AD$ (XVII, 29), and hence FG is level.

This form of level is very convenient for levelling large surfaces or long lines, as it can be made in any desired size. It was extensively used by surveyors in olden times, and is still so used when a modern surveyor's or carpenter's spirit level is not available. A level similar to this was used by the English surveyors Mason and Dixon in running the famous "Mason and Dixon Line" between Maryland and Pennsylvania.

Example 12. If the three bars AB, BC, CA, Fig. 71(a), are fastened together by pivots at A, B, C about which the bars are separately free to turn, the completed triangular framework cannot be distorted but remains rigid. This can be seen physically as follows. If the joint C is pressed in toward the link AB, in order that it may be movable either AB would have to stretch or the other two links would have to shorten, or both; and vice versa if C is pulled outward. Thus with rigid links C cannot move in either direction; and similarly for A and B. Thus with the links (sides) of *fixed lengths* the triangle cannot be distorted.

Geometrically this illustrates the principle of XI, 39, which states, in effect, that all triangles of equal sides are the same. That is, with three given, fixed sides, there can be only one form of triangle. This is sometimes expressed by saying that *a triangle is determined when its sides are given.*

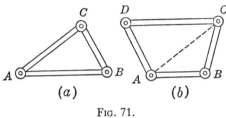

Fig. 71.

In Fig. 71(b), on the other hand, the joint C may freely be moved inward or outward and the links turn freely on their pivots without straining. If, however, a rigid link be connected across from A to C, as shown in the figure, then both ABC and ADC become rigid triangles as in (a). The same would happen if the connecting link were across BD. Thus a quadrilateral, as $ABCD$, *is not determined by its sides* alone. We shall see in the next section (B) the geometrical reason for this fact.

The principle of the rigidity of a triangle of fixed sides (XI, 39) is applied in most framework structures, such as bridges, trusses, building frames, swinging gates, furniture, machine frames, etc., by building them up of pieces so crossed and fastened as to form a network of rigid triangles.

Example 13. In Fig. 72 a pole or chimney AB is supported by stretched guy ropes or wires AC, AD, AE, etc., fastened to stakes set at C, D, E, etc., in the level ground. If AB is vertical and the wires are all of the same length, then the curved line on which the stakes are set must be a circle, with its center at B.

This is easily seen as follows. Since AB is vertical and the ground is level, AB is perpendicular to the ground and each of the angles ABC, ABD, ABE, etc., is a right angle. Hence each of the triangles ABC, ABD, ABE, etc., is a right triangle with the pole as a leg and the wire as hypotenuse. Since all the right triangles thus have one leg and

84 GEOMETRY FOR THE PRACTICAL WORKER CHAP. 4

the hypotenuse equal they are congruent (XXI, 41) and hence the corresponding legs BC, BD, BE, etc., are all equal.

Then since these distances are all equal, that is, the points C, D, E, etc., are all at the same distance from B, they are all on a circle and B is its center (article 14).

Example 14. It was seen in article 41 that the 45° right triangle is isosceles (has equal legs). This property provides a very simple and easily utilized method for measuring the height of tall objects such as poles, chimneys, trees, buildings, etc. This method is illustrated in Fig. 73, the height BC of the tree being desired.

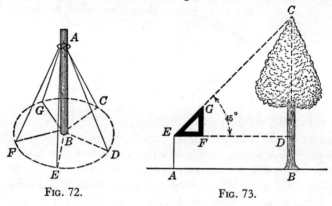

FIG. 72. FIG. 73.

The draftsman's 45° triangle EFG (or one cut from cardboard) is held in the hand with one leg (as EF) horizontal, and the other (FG) vertical, and with the eye at the vertex E. A position A is found where the top of the tree C is in the line of sight along the hypotenuse EG. The height BC of the tree is then equal to the distance AB plus the height of the person AE (to the eye).

For, in the 45° right $\triangle CDE$, the leg $CD = DE = AB$, and the height is $BC = CD + BD = AB + AE$.

44. Exercises.

1. Draw three triangles and with a protractor measure the three interior angles of each. Are the three sums of the angles of each triangle equal, and if so, what is each sum?
2. Can a right triangle be isosceles? If so, what is each acute angle?
3. Can a right triangle be equilateral? Explain.
4. Measure the three edges of a draftsman's 30–60° right triangle, and see whether the square of length of the longest side equals the sum of the

squares of the other two. How is the longest side related to the shortest?
5. If one interior angle of a certain triangle is 48°, what is the adjacent exterior angle?
6. What is each exterior angle of an equilateral triangle?
7. Draw a right triangle and extend one leg and the opposite end of the hypotenuse, forming two exterior angles. Show that each exterior angle equals 90° more than the opposite acute angle of the triangle.
8. Draw an isosceles triangle having the base angles each $72\frac{1}{2}°$, and the base six inches long. Measure the other two sides and see whether they are equal. What is the length of each? What is the altitude on the six-inch side?
9. Test proposition XII, 39, as follows: Lay off any angle and bisect it by means of the protractor; by means of the draftsman's triangle draw, from a point in the bisector, perpendiculars to both sides of the angle; measure these with a divided ruler.
10. Draw any triangle and by the protractor bisect each angle. Test proposition XIII, 40.
11. Draw triangles and test XIV, XV, XVI, 40.
12. The legs of a right triangle are three and six inches. What is the hypotenuse? Test by measurement.
13. The hypotenuse of a right triangle is 15 and one leg is 6. What is the other leg? Test by measurement.
14. Prove (as a theorem) that the bisectors of the base angles of an *isosceles* triangle are of equal length.
15. Prove that the bisector of an exterior angle of a triangle is perpendicular to the bisector of the adjacent interior angle.
16. Can a triangle have more than one obtuse angle? Explain.

B. Quadrilaterals

45. Forms and Descriptions of Quadrilaterals. A *quadrilateral* has been defined as a closed plane figure having four sides and four angles. As in the case of a triangle, when reference is made to *the angles* of a quadrilateral the interior angles will be meant, and the vertex of any one of these four angles is called a *vertex* of the quadrilateral. An *exterior* angle of a quadrilateral is the angle formed by extending a side through a vertex. *Adjacent* sides of a quadrilateral are any two sides which meet at a vertex. These several terms are illustrated in Fig. 74.

Thus in Fig. 74(a) the figure $ABCD$ is a quadrilateral with sides AB, BC, CD, DA, vertices A, B, C, D and angles ABC, BCD, CDA, DAB. The angle CBE, formed by extending AB through B to E, is an exterior angle, and similarly other exterior angles may be formed.

In the quadrilateral ABCD the vertices and angles A, C are *opposites*, as are B, D. Similarly AB and CD are opposite sides, as are AD and BC.

A quadrilateral which has no two sides parallel is called a *trapezium*. In Fig. 74, (a) represents a trapezium. A quadrilateral which has two and only two sides parallel is called a *trapezoid*. Figure 74(b) repre-

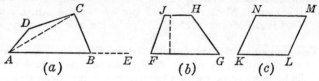

Fig. 74.

sents a trapezoid, having the sides FG and HJ parallel. A quadrilateral which has its opposite sides parallel in pairs is called a *parallelogram*, as (c) in Fig. 74. Here KL ∥ MN and KN ∥ LM. Parallelograms will be found to be the most important forms of quadrilaterals.

A parallelogram all of whose angles are *right* angles is called a *rectangle*. A rectangle all of whose *sides* are *equal* is called a *square*. Thus in Fig. 75, (a) is a rectangle and (b) a square.

A line joining opposite vertices of a quadrilateral is a *diagonal*, as AC in Fig. 74(a). A line joining BD would also be a diagonal, and similarly for any quadrilateral. Thus, any quadrilateral has two diagonals.

As in the case of a triangle, the side of a quadrilateral on which it is supposed to stand is its *base*. The two parallel sides of a trapezoid are called its two *bases*, regardless of the position in which the figure stands. In Fig. 74(b) the bases of the trapezoid FGHJ are FG and HJ. A line perpendicular to the bases of a trapezoid is called its *altitude*, as the dotted line in Fig. 74(b).

The side on which a parallelogram (or rectangle) stands and its opposite side are called its *lower* and *upper base*, as KL and MN, Fig. 74(c), or AB and CD, Fig. 75(a) and (c). Depending on the position

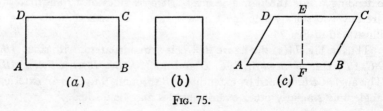

Fig. 75.

in which a parallelogram (or rectangle) stands, either pair of parallel sides may be taken as bases.

A line perpendicular to the bases of a parallelogram (either pair) is called its *altitude* on those sides, as *EF*, Fig. 75(c). A line perpendicular to *AD* and *BC* would be the altitude on those sides. In the case of a rectangle, when any one pair of sides is taken as bases, the other pair, being perpendicular (at right angles) to these two, are altitudes. Thus, in Fig. 75(a) *AB*, *CD* are bases and *AD*, *BC* are altitudes.

In the case of an oblique parallelogram (not a rectangle), such as Fig. 75(c), the altitude must not be confused with a side. Thus *AD* and *BC* reach from one base to the other, but they are not *perpendicular* to the bases, as is the altitude *EF*.

In the same way as the equilateral triangle is unique among triangles, the square, which is the equilateral and equiangular quadrilateral, is also unique among quadrilaterals. It possesses many properties possessed by no other quadrilateral.

In the next article we study some of the properties of quadrilaterals, which will be found to be related to those of triangles.

46. Some Properties of Quadrilaterals. A square is a form of parallelogram and its sides are equal by definition. In general, however, the sides of a parallelogram are only defined to be *parallel*. The question at once arises, therefore, are any of the sides of a parallelogram *equal?* This question is answered by the following easily proved theorem:

XXVIII. *The opposite sides of a parallelogram are equal.*

In Fig. 76 let *ABCD* represent any parallelogram, and draw a diagonal, say *AC*. Then by definition of a parallelogram, *AB*, *CD* are parallels, and so are *AD*, *BC*; and *AC* is a transversal cutting both pairs of parallels. Therefore the alternate-interior $\angle BAC = \angle DCA$, and also $\angle BCA = \angle DAC$ (XVIII, 31). Hence, in the two triangles *ABC*, *ADC* the

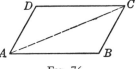

Fig. 76.

two angles *BAC*, *BCA* and included side *AC* are equal to the two angles *DCA*, *DAC* and included side *CA*, and therefore the two triangles are congruent (V, 39). Therefore the corresponding parts $AB = CD$, and also $AD = BC$, as was to be proved.

In the course of the proof just given it was found that the two triangles *ABC*, *ADC* are congruent; a similar result would be obtained if the diagonal *BD* were drawn. In general, therefore,

XXIX. *A diagonal divides a parallelogram into two congruent triangles.*

Suppose a parallelogram had not been defined by the parallelism of its sides, and the theorem XXVIII had not been proved. If then it were simply stated that a certain quadrilateral has its opposite sides *equal*, could it be concluded at once that they are also *parallel?* This question cannot be answered without constructing and testing such a quadrilateral, or reasoning the matter out.

Thus, suppose that in Fig. 76 it is known that the opposite sides are equal in pairs, but no more is known, and suppose the diagonal drawn. Then by hypothesis, $AB = CD$, $AD = BC$, and as an identity $AC = CA$. That is, the three sides of $\triangle ABC$ equal the three sides of $\triangle CDA$. Hence the two triangles are congruent (XI, 39), and the corresponding parts $\angle BAC = \angle DCA$. Since the alternate-interior angles of the transversal AC are equal, therefore $AB \parallel CD$. In the same way we find that also $AD \parallel BC$. That is, the opposite sides are parallel.

Our original question is therefore answered as follows:

XXX. *If the opposite sides of a quadrilateral are equal the figure is a parallelogram.*

It is thus not necessary to have given the equality *and* the parallelism of opposite sides of a quadrilateral in order to determine that it is a parallelogram, therefore, but only the parallelism *or* the equality alone. If either is given the other is a necessary consequence.

Similarly, it is not necessary to have given the equality and parallelism of *both* pairs of opposite sides of a quadrilateral in order to prove it a parallelogram. If both the parallelism and the equality are given it is only necessary to specify these relations for *one* pair of sides. That is,

XXXI. *If two sides of a quadrilateral are equal and parallel then the other two are also equal and parallel,*

which of course makes the figure a parallelogram, according to the last two theorems. The proof of this proposition is easy and is left as an exercise for the reader.

The three propositions XXVIII, XXX, XXXI form a complete trio, and should be particularly noted. Thus, since by definition a parallelogram has its opposite sides parallel, XXVIII really states that if *both* pairs of opposite sides are *parallel* they are also equal; XXX states independently that if *both* pairs are *equal* they are also parallel; and XXXI states, also independently, that if *one* pair are both equal *and* parallel, so also are the other pair.

Thus *two conditions* involving the sides are necessary to determine completely that a quadrilateral is a parallelogram, and there are three pairs of such conditions: *two* pairs of sides equal, *two* pairs parallel; *one* pair equal *and* parallel.

When not only the sides but also the angles are involved, a parallelogram is determined in another way; stated in the form used in connection with congruent triangles this is:

XXXII. *If two sides and the included angle of one parallelogram are equal to two sides and the included angle of another, the two parallelograms are congruent,*

or, in other words, a parallelogram is also determined by two sides and their included angle. Thus, if we consider *opposite* sides they must be equal and parallel; if we consider *adjacent* sides they must be respectively equal and form the same angle.

This is seen from Fig. 76 without formal proof. Thus consider the two sides AB and CB and their included $\angle ABC$; since the other sides must be parallel to these to form a parallelogram, they can each be drawn in only one position (parallel axiom) and so can form only one complete figure.

The following theorem states a remarkable property of a parallelogram:

XXXIII. *The diagonals of a parallelogram bisect each other.*

In order to show this to be so, let $ABCD$, Fig. 77, represent any parallelogram, having diagonals AC, BD intersecting at O. We have to prove that O is the midpoint of both the diagonals; that is, that $OA = OC$ and $OB = OD$.

Mark the angles a, b, c, d formed by the diagonals with the bases, as shown. Then from the alternate-interior angle theorem on transversals, we have, since the opposite sides are parallel, $\angle a = \angle c$ and $\angle b = \angle d$. Also the opposite sides AB, CD are equal (by XXVIII).

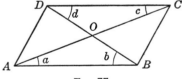

Fig. 77.

Therefore the two ∠a, b and included side AB of △ABO are equal to the two ∠c, d and included side CD of △CDO, and hence the two triangles are congruent. Therefore, as corresponding parts, OA = OC and OB = OD, as was to be proved.

We next prove by the use of properties of parallelograms, an important proposition concerning parallels and transversals. By association, this proposition belongs in article 31 but the proof was there postponed and it is placed here on account of its dependence on the parallelogram theorems:

XXXIV. *If three or more parallels cut off equal segments on one transversal they do so on every transversal.*

In Fig. 78 let AB, CD, EF, GH be a set of parallels which cut off equal segments AC, CE, EG on the transversal AG, and let BH be any other transversal of the same parallels, on which the parallels cut off segments BD, DF, FH. We have to prove that BD = DF = FH.

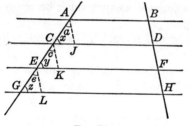
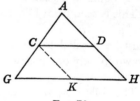

Fig. 78.   Fig. 79.

Draw AJ, CK, EL all parallel to BH, and hence parallel to each other (XV, 31). Then the corresponding angles a, c, e formed by these parallels and the transversal AG are all equal, as are the corresponding angles x, y, z formed by the transversal AG and the given parallels (XX, 31). Also the original segments AC, CE, EG are equal by hypothesis. That is, two angles and the included side are the same in each of the triangles ACJ, CEK, EGL, and hence the triangles are congruent; and hence the corresponding sides AJ, CK, EL are all equal.

Now the quadrilaterals ABDJ, CDFK, EFHL have their long sides parallel by hypothesis and their ends parallel by construction. They are therefore, by definition, parallelograms, and hence the ends BD = AJ, DF = CK, FH = EL. But we have already proved AJ = CK =

EL. Therefore, by the equality axiom, $BD = DF = FH$, as was to be proved.

This result can be used to bring out some very interesting properties of triangles. Thus suppose that the transversal BH passes through A, and that there are only two parallels, as in Fig. 79. The figure then becomes the $\triangle AGH$, with one of the parallels GH as base and the other CD as a line parallel to the base and cutting the other two sides of the triangle. If then $AC = CG$, we have at once also $AD = DH$. Therefore, in general,

XXXV. *If a line parallel to the base of a triangle bisects one side it bisects the other side also.*

Suppose now, without regard to XXXV, that CD is drawn originally to join the middle points C, D of the sides AG, AH of the triangle AGH. Then, even if we did not have XXXV we could find as in XXXIV that CD is parallel to the base.

We can now find another relation between CD and GH. For this purpose draw $CK \parallel AH$. Then, since CK bisects AG and is parallel to AH, it also bisects GH (XXXV). That is, $KH = \frac{1}{2}\overline{GH}$.

Also, since we have found that $CD \parallel KH$ and also drew $CK \parallel DH$, $CDHK$ is a parallelogram, and therefore side $CD = KH$. But $KH = \frac{1}{2}\overline{GH}$; therefore, $CD = \frac{1}{2}\overline{GH}$, by the equality axiom. That is, CD, which bisects both AG and AH, is *parallel to* GH and *equals half* GH. In general, therefore,

XXXVI. *The line which joins the middle points of two sides of a triangle is parallel to the third side and equal to half of it.*

From this truly remarkable result pertaining to the triangle we can obtain also an interesting and remarkable property of the form of quadrilateral which we have called a *trapezoid*. We first give a definition:

The line joining the middle points of the non-parallel sides of a trapezoid is called the *median* of the trapezoid.

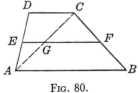

Fig. 80.

Consider now the trapezoid $ABCD$ in Fig. 80, with EF as the median, and AB, CD (parallels) as the bases. The points E, F being the middle points of AD, BC then $AE = ED$ and $BF = FC$; these segments of the transversals AD, BC being equal, the lines AB, CD,

EF are therefore parallel (XXXIV). Thus the *median is parallel to the bases*.

Now draw the diagonal *AC*, intersecting the median at *F* and dividing the trapezoid into the two triangles *ABC*, *ADC*. The median *EF* is now parallel to the sides *DC*, *AB* of the triangles and by definition bisects the sides *AD*, *BC*. According to XXXV, therefore, it bisects the side *AC* of both triangles. Similarly it would *bisect the diagonal BD* if it were drawn.

Furthermore, according to XXXVI, in the $\triangle ABC$,

$$GF = \tfrac{1}{2}\overline{AB}$$
and in $\triangle ADC$, $$EG = \tfrac{1}{2}\overline{CD}$$
Adding, $$EG + GF = \tfrac{1}{2}\overline{AB} + \tfrac{1}{2}\overline{CD}$$
$$= \tfrac{1}{2}(AB + CD)$$

But $EG + GF = EF$, the median; and $AB + CD$ is the sum of the bases of the trapezoid. Therefore, the *median equals half the sum of the bases*.

Summing up the three results just obtained, we have the following:

XXXVII. *The median of a trapezoid bisects the diagonals and is parallel to the bases and equal to half their sum.*

We demonstrate next a remarkable property possessed by all quadrilaterals in common, regardless of their shape. It is:

XXXVIII. *The lines joining the middle points of adjacent sides of any quadrilateral form a parallelogram.*

In Fig. 81 let *ABCD* represent any quadrilateral; *E*, *F*, *G*, *H* the middle points of its sides; and *EF*, *FG*, *GH*, *HE* the lines joining adjacent middle points. To prove: that *EFGH* is a parallelogram.

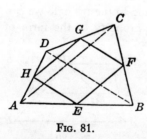

Fig. 81.

Draw the diagonals *AC*, *BD* of the quadrilateral, forming four triangles. In $\triangle ABC$, *EF* joins the middle points of sides *AB*, *BC*, by hypothesis, and hence *EF* is parallel to the third side *AC* (according to XXXVI). Similarly in $\triangle ADC$, $GH \parallel AC$; and hence $EF \parallel GH$, as both are parallel to the same line. In the same way we find, in triangles *BAD*, *BCD*, that $FG \parallel HE$.

That is, the opposite sides of the quadrilateral *EFGH* are parallel, and hence it is a parallelogram, as was to be proved.

Furthermore, in $\triangle ABC$, $EF = \frac{1}{2}\overline{AC}$, and in $\triangle BCD$, $FG = \frac{1}{2}\overline{BD}$ (according to XXXVI). Then, since in the parallelogram *EFGH*, the opposite sides $GH = EF$ and $HE = FG$, $EF + GH = AC$ and $FG + HE = BD$. Adding these last two equations $(EF + FG + GH + HE) = (AC + BD)$. But the sum on the left is the sum of the sides, or *perimeter*, of *EFGH*, and that on the right is the *sum of the diagonals* of the original quadrilateral *ABCD*. Therefore,

XXXIX. *The perimeter of the parallelogram formed by lines joining the midpoints of adjacent sides of any quadrilateral equals the sum of the diagonals of the quadrilateral.*

We are now in position to prove the propositions XV, XVI of article 40, concerning the altitudes and medians of triangles.

XV. *The three altitudes of a triangle meet in one point.*

In Fig. 82(*a*) below let *ABC* be the triangle and *AD*, *BE*, *CF* the altitudes, each perpendicular to a side, by definition. The altitudes are shown as meeting at *O*; this is to be proved.

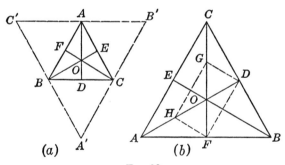

Fig. 82.

Through the vertices *A*, *B*, *C* draw $B'C'$, $C'A'$, $A'B'$ parallel to the opposite sides. Then $AD \perp B'C'$, because it is $\perp BC$ (XVII, 31); and similarly $BE \perp A'C'$, $CF \perp A'B'$.

Also, since the sides of $\triangle A'B'C'$ are drawn parallel to those of $\triangle ABC$, the figures $ABCB'$ and $ACBC'$ are parallelograms, and hence the opposite sides $AB' = BC$, $AC' = BC$ (XXVIII, 46). Hence $AB' = AC'$

and A is the midpoint of $B'C'$. Similarly B, C are the midpoints of $A'C'$, $A'B'$.

We have thus proved that the altitudes of the original $\triangle ABC$ are perpendicular to the sides of the $\triangle A'B'C'$ at their midpoints. That is, AD, BE, CF are the perpendicular bisectors of the sides of $\triangle A'B'C'$. Therefore they meet in one point (XIV, 40).

XVI. *The three medians of a triangle meet in one point, whose distance from each vertex is two-thirds the median from that vertex.*

In Fig. 82(b) let ABC be the triangle, with AD, BE, CF as medians joining the vertices A, B, C to the midpoints of the opposite sides (by definition of median). The medians are shown passing through the same point O. We are to prove that this is so, and also that $AO = \frac{2}{3}\overline{AD}$, $BO = \frac{2}{3}\overline{BE}$, $CO = \frac{2}{3}\overline{CF}$.

Since the medians are not parallel, certainly any two of them intersect, say AD and CF at O. Take G the midpoint of CO and H of AO. Join DF, GH, DG, FH.

Then in $\triangle AOC$, GH joins the midpoints of two sides and hence $GH \parallel AC$ the third side and also $GH = \frac{1}{2}\overline{AC}$ (XXXVI, 46). Similarly DF joins the midpoints of two sides of $\triangle ABC$ and is parallel to and equals $\frac{1}{2}\overline{AC}$. Hence DF $\parallel$ GH and $DF = GH$, and therefore $DFGH$ is a parallelogram (XXXI, 46), and hence also the diagonals DH, FG bisect one another. That is, $HO = OD$ and $GO = OF$ (XXXIII).

But G and H were so taken that $AH = HO$ and $CG = GO$. Therefore $AH = HO = OD$ and $CG = GO = OF$. From this $AO = \frac{2}{3}\overline{AD}$ and $CO = \frac{2}{3}\overline{CF}$, as was to be proved.

Now AD, CF were taken as *any* two of the three medians. Hence if we take AD, BE then in the same way we find $BO = \frac{2}{3}\overline{BE}$, which completes the proof.

47. Quadrilateral Constructions and Applications. In this article we combine a few constructions and applications involving quadrilaterals such as are given in articles 42 and 43 for triangles.

(i) *To divide any given line into any number of equal parts.*

This is an operation which is a regular part of the work of a draftsman. Of course it can be done approximately by means of a graduated ruler, but the following exact method is simpler. It is based on the proposition XXXIV of the preceding article.

Let AB, Fig. 83, be the given line (length not measured) and let it be required to divide it, say, into five equal parts.

Through A draw AC, at any convenient angle and of any convenient length. With the dividers step off from A on AC *five* equal segments of any convenient length, marking the points 1, 2, 3, 4, 5, and join the fifth point to B by the line $B5$. Through the other points of division of AC draw parallels to $B5$, intersecting AB. (See Fig. 38, article 33.) These parallels divide AB into five equal parts.

For the parallels cut off equal segments on the transversal AC, and hence also on AB, according to XXXIV, 46.

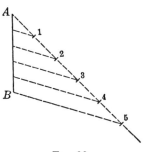

Fig. 83.

(ii) *A "parallel ruler" or device for drawing parallel lines.*

In Fig. 84(a) AB and CD are two straight-edged rulers fastened together by the equal cross pieces EF and GH, pivoted at E, F, G and H so that $EG = FH$. If AB is held stationary on the drawing board while CD is moved, then CD will in any position be parallel to AB and its edge may be used as a ruler to draw straight parallel lines.

This is seen at once to be true because the opposite sides of $EFGH$, being equal, form a parallelogram (XXVIII, 46) and hence, the opposite sides are parallel.

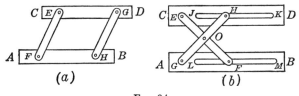

Fig. 84.

In Fig. 84(b) AB and CD are two straight rulers, slotted as shown at JK and LM. EF and GH are two equal cross bars pivoted on AB and CD at G and E, and pivoted together at O at the center of each. If AB is held stationary on the drawing board while CD is moved, CD will in any position be parallel to AB and may be used as a ruler to draw parallels in such positions.

For, as H and F slide in the slots JK and LM, EF and GH being permanently fastened at their common midpoint O, the point O always bisects EF and GH. EF and GH thus bisecting each other and being the diagonals of the quadrilateral $EFGH$, the quadrilateral is a parallelogram (XXXIII, 46), that is, $CD \parallel AB$.

(iii) *To determine the distance between two inaccessible points.*

The methods of measuring an inaccessible line by means of triangles, given in article 43, require that one of the ends of the line be accessible.

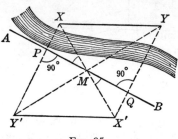

Fig. 85.

By means of quadrilaterals, however, such a line may be measured when neither end is accessible.

Thus suppose that the points X, Y, Fig. 85, are visible objects beyond a river and it is desired to find their distance apart without crossing the river.

Run a line AB along the bank of the river. Then take a carpenter's square or a large square-cornered book or card, and walk along AB until a point P is reached from which X and B can be seen along the edges of the square, thus making $\angle XPB$ a right angle. Similarly locate the point Q to make $\angle YQA$ a right angle; and set stakes at P and Q.

Set a stake at the middle point M of PQ and walk back along YM extended, until a point Y' is reached from which P and X are in line; and on XM extended similarly locate X'. Then the distance $X'Y' = XY$ and can be measured at once.

This is readily seen to be true, as follows. Since the angles at P and Q are right angles, MPX and MQX' are right triangles; and since the vertical $\angle PMX = \angle QMX'$ and $PM = QM$, the two triangles are congruent (XIX, 41). Hence the corresponding parts $MX' = MX$. Similarly in the triangles MPY', MQY, $MY' = MY$.

Therefore, M is the middle point of the diagonals XX' and YY', and hence $XYX'Y'$ is a parallelogram (XXXIII, 46). Therefore, as opposite sides, $X'Y' = XY$.

48. Exercises.

1. Cut a parallelogram from paper, and cut it in two parts along a diagonal. See whether the two resulting triangles are alike and of the same size.

2. On a sheet of ruled writing paper draw any two diagonal straight lines, and by measurement test proposition XXXIV, 46.

3. Draw a parallelogram and its two diagonals and by measurement test XXXIII, 46.

4. A triangle ABC has sides $AB = 10$, $BC = 6$, $AC = 8$ inches. A line DE is drawn parallel to AB and cutting the other two sides at D, E. If D is the middle point of AC, how long are BE, CE, DE? Draw the figure.

5. Draw carefully a figure like Fig. 80 ($AB \parallel DC$) and by measurement test XXXVII, 46.

6. If in Fig. 81 the diagonals are $AC = 12$, $BD = 8$ inches, what are the sides of $EFGH$? Can the shape of $ABCD$ (that is, the sides) change without changing the diagonals? If these sides are so changed, will the sides of the parallelogram be changed?

7. Prove as a theorem that the lines joining the midpoints of the sides of a triangle divide it into four congruent triangles.

8. Prove that the bisectors of the two opposite angles of a parallelogram are parallel.

C. Polygons

49. Definitions and Descriptions. A rectilinear closed plane figure of any number of sides is called a *polygon* (see article 36). Polygons are designated according to the number of sides or angles, as stated in article 36.

A few of the names of polygons are listed here as definitions.

No. of Sides	Name of Polygon
3	triangle
4	quadrilateral
5	pentagon
6	hexagon
7	heptagon
8	octagon
9	nonagon
10	decagon
11	undecagon
12	dodecagon
15	pentadecagon

Polygons of thirteen, fourteen and more than fifteen sides are designated by stating the number of sides, as "a polygon of 13 sides" or as "a 13-sided polygon." For simplicity this is sometimes written as 13-*gon*. Similarly a polygon of any number (n) of sides is sometimes called an *n-gon*.

In sections A and B of this chapter we have already considered some of the properties of triangles and quadrilaterals; in this section we consider a few of the properties which apply to polygons in general, including those of three and four sides.

Fig. 86 shows polygons of several forms.

In this figure (*a*) represents a *pentagon*, (*b*) a *hexagon*, (*c*) a *heptagon*, and (*d*) a *heptagon* also. The angle at the vertex 3 in (*d*) is called a *re-entrant* angle, and a polygon having one or more such angles is called a *concave* polygon. All other polygons, having no re-entrant angles, are said to be *convex*. Thus Fig. 86(*c*) is a convex heptagon, (*d*) is a concave heptagon.

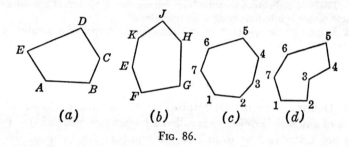

FIG. 86.

In this book all polygons will be understood to be convex unless otherwise specified.

A polygon which has all its *sides* of equal length is called an *equilateral* polygon; one which has all its *angles* equal is said to be *equiangular*. A polygon may be equilateral without being equiangular, and vice versa, except in the case of a triangle.

A polygon which is *both* equilateral *and* equiangular is said to be a *regular* polygon.

In Fig. 87, (*a*) is an equilateral hexagon, (*b*) is an equiangular hexagon, and (*c*) is a regular hexagon.

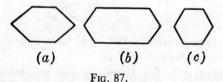

FIG. 87.

Two polygons whose corresponding sides or corresponding angles are equal each to each are said to be *mutually* equilateral or equiangular, respectively. Two polygons which are mutually equilateral

ART. 50 PLANE FIGURES 99

and equiangular are congruent. Two congruent polygons when superposed with corresponding parts together will coincide, part by part, throughout.

The equilateral triangle is a regular trigon; the square is a regular quadrilateral; the locking nut or tap on some ordinary street hydrants (fire plugs) is an example of a regular pentagon; an ordinary six-faced machine screw nut is a familiar example of a regular hexagon, another is the cross section (end view) of many familiar writing and drawing pencils.

The vertex of an angle formed by two adjacent sides of a polygon is a *vertex* of the polygon (as already used in the case of the triangle and quadrilateral); the angle inside the polygon formed by the two sides at a vertex is an *interior* angle of the polygon; and the angle formed outside the polygon by extending one side through a vertex is an *exterior* angle of the polygon.

A line other than a side joining any two vertices of a polygon is called a *diagonal*.

The sum of the sides of a polygon is called the *perimeter* of the polygon.

In the next article we investigate a few very important properties of polygons in general and show how these apply in particular to triangles and quadrilaterals.

50. Properties of Polygons. According to proposition I, 39, the sum of the interior angles of a triangle is one straight angle (180°). This is the fundamental theorem on which are based all similar theorems for other polygons. This question was not investigated in connection with quadrilaterals. If we can get a general statement concerning the angles of *any* polygon (and hence *all* polygons) this will of course include the quadrilateral, as well as the pentagon, etc.

The general relation for any polygon is

XL. *The sum of the interior angles of any polygon is a number of straight angles which is two less than the number of sides.*

Let the polygons of Fig. 88 represent polygons of any number of sides; thus that in (*a*) has five sides and that in (*b*) has six. In general let $ABCD\ldots$, etc., be a polygon of n sides. We have to prove $\angle A + \angle B + \angle C + \angle D + \ldots = (n-2)$ st. $\angle$.

From any vertex of either polygon draw diagonals to all the other vertices, forming two triangles less than the number of sides. Then in the pentagon there are **three** such triangles, in the hexagon there are four, in the heptagon five, etc., and in the n-gon there are $(n-2)$ triangles.

In Fig. 88(a), for example, the $\angle E$ is an angle of the polygon and also of $\triangle ADE$, and similarly for $\angle C$ and $\triangle DCB$. Also the polygon angle at D is the sum of the three angles of the three triangles (one in each triangle) whose vertices are formed at D by the diagonals; and at A and B the polygon angle is the sum of two triangle angles. Thus the sum of all the angles of the three triangles is also the sum of the interior angles of the polygon. The same is true of the hexagon, Fig. 88(b), and similarly of any n-gon.

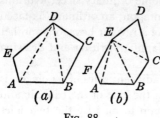

Fig. 88.

But the sum of the angles of each triangle is 1 st. $\angle$ (I, 39), and hence the sum of the $(n-2)$ triangles in the polygon is $(n-2)$ straight angles. That is, for any polygon $ABCD\ldots$ of n sides,

$$\angle A + \angle B + \angle C + \angle D + \ldots = (n-2) \text{ st. } \angle\!\!\!\!\angle.$$

Since a straight angle equals 180°, this proposition is very simply stated in the following form:

XLI. *The sum of the interior angles of a polygon of n sides is $(n-2)$ 180 degrees.*

Since a quadrilateral has four sides, the sum of the interior angles of any quadrilateral is two straight angles, or since $n = 4$, the sum is $(4 - 2)\, 180 = 2 \times 180 = 360°$. Thus,

XLII. *The sum of the interior angles of any quadrilateral is two straight angles, or four right angles, or 360 degrees.*

If a polygon is equiangular all its n angles are equal, and each is $\frac{1}{n}$th of $(n-2)$ straight angles, or $\left(\frac{n-2}{n}\right)180°$. But, $\frac{n-2}{n} = 1 - \frac{2}{n}$ (by algebra). Hence we can say that

XLIII. *Each interior angle of a regular polygon of n sides is $\left(1 - \frac{2}{n}\right)180$ degrees.*

As an illustration, consider the square (regular quadrilateral). The sum of the interior angles being two straight angles or four right angles,

Art. 50   PLANE FIGURES   101

each angle is one right angle or $360 \div 4 = 90°$. By the formula, $n = 4$, $(1 - \frac{2}{4}) = 1 - \frac{1}{2} = \frac{1}{2}$, and hence $(1 - \frac{2}{4})180 = \frac{1}{2} \times 180 = 90°$.

Similarly, each interior angle of a regular pentagon ($n = 5$) is $(1 - \frac{2}{5})180 = \frac{3}{5} \times 180 = 108°$. The sum of the angles is $5 \times 108 = 540°$, or directly by the original formula, $(n - 2)180 = 3 \times 180 = 540°$.

In the same way the sum, and the amount of each, of the interior angles of any regular polygon can be found, or the sum of the interior angles of any polygon whether regular or not.

The next question which arises is: What is the sum of the *exterior* angles of any polygon? The answer to this question is:

XLIV. *The sum of the exterior angles of any polygon is two straight angles or 360 degrees.*

Let Fig. 89 represent a polygon of any number of sides, say n, and angles A, B, C, D . . . , with all the sides produced in the same order going around the polygon, to form the exterior angles a, b, c, d . . . . To prove $\angle a + \angle b + \angle c + \angle d + \ldots = 2$ st. $\angle$.

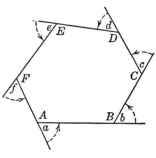

Fig. 89.

At each vertex the adjacent interior and exterior angles, as B and b, form a straight angle. Hence, if there are n vertices the sum of all the interior *and* exterior angles is n straight angles. That is,

$$A + a + B + b + C + c + D + d + \ldots = n \text{ st. } \angle$$
or, $(A + B + C + D + \ldots) + (a + b + c + d + \ldots) = n \text{ st. } \angle$
But $(A + B + C + D + \ldots) \qquad\qquad\qquad\qquad = n - 2$
Subtracting, $\qquad\qquad (a + b + c + d + \ldots) = 2 \text{ st. } \angle,$

as was to be proved.

It sometimes seems strange to beginners that the sum of the interior angles should depend on the number of sides while the sum of the exterior angles does not, when there is one interior and one exterior angle at each vertex.

The following considerations are sometimes brought forward to clear up the question. Suppose a single line, say AB in the figure, is thought of as being used to form each side of the polygon in succession, by sliding it to the right until A reaches B, and then turning it through $\angle b$ to the position BC, as shown by the dotted arrow; and repeat the process at each vertex. When the original position AB is again reached the line has turned through one complete revolution, or 360°, and has also turned through all the exterior angles, a, b, c, d, etc. The sum of these angles is therefore the complete revolution, 360°, or two straight angles, and cannot be more or less than 360° because in turning through them the generating line completes only one rotation to return to its original position.

The sum of the exterior and interior angles at each vertex is, however, one straight angle, and there are as many of these as there are vertices. So the total of interior and exterior angles depends directly on the number of vertices, or sides. Subtracting the fixed sum of the exterior angles, there remains the (variable) sum of the interior angles, which depends therefore on the number of sides.

The properties of polygons developed in this article apply to all polygons. Regular polygons (such as the regular quadrilateral or square and the regular trigon or equilateral triangle) possess a number of special properties of great interest and importance, but they will be treated in a later chapter in connection with the circle.

We may use the fact that the sum of the exterior angles of a polygon is 360 degrees to solve a variety of problems as exemplified below.

Example 1. If an interior angle of a regular polygon is 108° find the number of sides of the polygon.

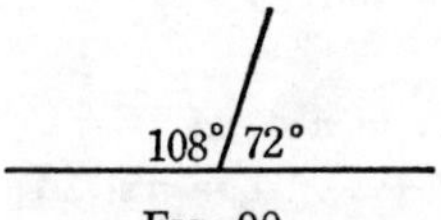

Solution. Since an interior angle of a polygon is supplementary to its corresponding exterior angle, the exterior angle is 72° (as shown in the diagram).

Fig. 90.

The sum of all the exterior angles of a polygon is 360°. Therefore, we may find the number of sides of the polygon by dividing 360 by 72. The result is 5. The polygon has 5 sides.

Example 2. A regular polygon has 24 sides. Find the number of degrees in one of its interior angles.

Solution. The sum of all the exterior angles of a polygon is 360°. We may find the number of degrees in each exterior angle of a regular polygon of 24 sides by dividing 360 by 24. Thus, each exterior angle of a polygon of 24 sides contains 15°.

Since each exterior angle of a polygon is supplementary to its corresponding interior angle, the number of degrees contained in each interior angle of this polygon is 180° − 15° = 165°.

51. Exercise.

Calculate the sum of the interior angles of each polygon of the list in article 49. If these are regular polygons, what is the value of each interior angle?

Chapter 5

SOME PROPERTIES OF THE CIRCLE

52. The Circle. In the preceding chapter we studied plane figures formed of straight lines alone. In the present chapter we take up the study of plane figures formed in part or entirely of curved lines. A curved line has already been defined as a line no part of which is straight (article 15). The particular curved line or lines which we shall study is the *circle* or parts of a circle. A part of a circle is called a circular *arc*.

The circle is a familiar figure, and has been defined in article 14 in connection with the compass. The compass and a few properties of the circle have already been used in certain constructions and in discussing angle measure (articles 26, 27). In this chapter we shall consider in detail many of its important properties and its relations to certain straight lines. For this purpose we re-state here some of the definitions of article 14.

A CIRCLE *is a closed plane figure formed by a curved line every point of which is equally distant from one and the same point inside the figure.*

The *length* of the curved line which forms the circle is called the *circumference* of the circle. (This term is also used sometimes simply to name the curved line without reference to its length, but the word "circle" is the *name* of the curved line.)

The point inside the circle which is equally distant from every point of the circle is called the *center* of the circle.

A straight line joining the center to any point on the circle is called a *radius* of the circle.

A straight line passing through the center and having its two ends on the circle is called a *diameter* of the circle.

The terms "radius" and "diameter" are also used to mean the lengths of the lines having these names. In this case the length is usually referred to as *the radius* or *the diameter*.

When a circle or an arc of a circle is drawn to satisfy any specified conditions it is said to be *described*.

We recall and state here the *postulate* (No. 13) already given in article 21:

A circle may be described with any given point as center and with any given radius.

From the definition of a circle the following statement is at once seen to be true:

Two circles are equal if their radii are equal.

For, described with the same radius, they will coincide if their centers are placed at the same point, and equal figures are those which can be made to coincide throughout.

Two or more circles having the same center, or their centers at the same point, are said to be *concentric*.

If a smaller circle is drawn inside a larger one but the two have different centers, they are said to be *eccentric* (pronounced "ekcentric").

A circle is generally designated in writing and printing by means of the symbol ⊙ and a letter indicating the point which is its center. Thus ⊙C, which is read "the circle with center at C" or simply "circle C"; similarly ⊙O, "circle O." If it is necessary to describe the circle in more detail the radius or diameter is also stated, or at least three points through which it passes are named.

In Fig. 91 is shown the ⊙O with radius OP and diameter AB. This circle may also be described as "the circle through A, P, B with center O" or simply as "circle APB."

The algebraic symbols representing the lengths of radius, diameter and circumference are R, D and C, respectively. In

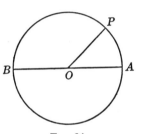

Fig. 91.

Fig. 91 it is obvious, from the definitions, that OA and OB as well as OP, are radii (plural of "radius") of the circle, and hence that AB is twice the length of OA, OB or OP. Thus $D = 2R$, or $R = \frac{1}{2}D$. We shall find later the relations between C and R or D.

In the next article we give the chief definitions which are used in describing the various properties and characteristics of the circle and its relations to certain associated straight lines.

53. Definitions Pertaining to the Circle. A straight line of any length which intersects a circle at any two points is called a *secant*, as AB, Fig. 92(a), intersecting the circle at C, D. The portion of the circle between C and D is the *arc CD*. Also in Fig. 91 the portions AP and PB are arcs. The expression "arc CD" is written in symbols as $\overset{\frown}{CD}$.

Any straight-line segment which has its ends on a circle is called a *chord* (pronounced as "cord" but not to be confused with this word), as EF, Fig. 92(a), or AB in (b). Obviously a diameter is also a chord.

The secant AB and the chord EF in (a) are said to *intercept* the arcs CD, EF, respectively, and the arc EF is said to *subtend* the chord EF, in (a).

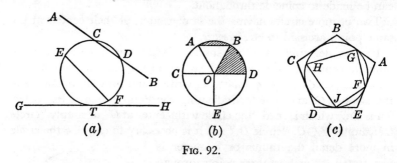

Fig. 92.

A straight line of any length which touches a circle at one and only one point is called a *tangent*. It is said to be "tangent *to* the circle *at* the point of *contact*"; this point is also called the point of *tangency*. In Fig. 92(a) GH is tangent to the circle at T.

The figure formed by an arc of a circle and its subtended chord is called a *segment* of the circle, as the shaded figure AB in Fig. 92(b).

The angle whose vertex is at the center of a circle and whose sides are radii of the circle extending to the ends of a chord, is called the angle of the chord, and is also said to be *subtended* by the chord, or by the arc which the chord intercepts. Thus, in Fig. 92(b), $\angle AOB$ is the angle of the chord AB and is subtended by chord AB or arc AB.

Any angle whose vertex is at the center of a circle and whose sides are radii of the circle is also called a *central angle* of the circle. Thus in (b) the angles AOC, AOB, BOD, etc., are central angles.

The arc cut off by or included between the sides of a central angle is said to be *intercepted* by the central angle, and the angle is said to be *subtended* by the arc (as above). Thus in (b) central $\angle AOC$ intercepts $\overset{\frown}{AC}$ and $\overset{\frown}{AC}$ subtends $\angle AOC$.

The figure formed by an arc of a circle and its subtended central angle is called a *sector* of the circle, as the shaded figure *BOD* in (*b*).

A sector equal to a fourth or quarter of a circle is called a *quadrant*, as *COE* or *DOE* in (*b*). A quadrant is thus formed by a 90° or right central angle (see article 27).

A segment formed by a diameter, that is, half the circle, is called a *semi-circle*, as *CED* in (*b*).

The arcs subtending a quadrant or semi-circle are also called by the same names.

A polygon is said to be *inscribed in a circle* if its sides are chords, or if all its vertices are on the circle; the circle is then said to be *circumscribed about* the polygon.

A polygon is said to be *circumscribed about a circle* if its sides are tangents; the circle is then said to be *inscribed* in the polygon.

In Fig. 92(*c*) the pentagon *ABCDE* is circumscribed about the circle, and the quadrilateral *FGHJ* is inscribed in the circle. The circle is inscribed in *ABCDE* and circumscribed about *FGHJ*.

If the vertex of an angle is on a circle and the sides of the angle are chords of the circle, the angle is called an *inscribed angle* of the circle, as ∠*GHJ* in (*c*). (An *inscribed* angle of a circle is not to be confused with a *central* angle.)

We now proceed to discuss some of the relations and properties of the lines described above and their relations to the circle.

54. Properties of Arcs, Chords, and Tangents. We have already seen that two straight lines can intersect in only one point (II, 22). It is also obvious that

I. *A straight line and a circle cannot intersect in more than two points.*

For if they could then three or more points of the circle would lie on the straight line; but from the definitions of the straight line and circle (curve) this is impossible.

We shall investigate later the intersection of *two circles*.

We state here two properties of circles which are obvious without formal proof.

II. *In the same or equal circles, equal central angles intercept equal arcs, and equal arcs subtend equal central angles.*

III. *In the same or equal circles, equal chords intercept equal arcs, and equal arcs subtend equal chords.*

The first of these means simply that if two equal sectors are cut out of the same circle (or equal circles) the curved edges will be equal; and the second means that if the chords are then drawn joining the two ends of each arc these chords will be equal. This is seen at once by considering the two triangles formed by the sector radii and the chords. They have two sides (the radii) and included (central) angle equal and hence (IV, 39) the third sides (chords) must be equal.

IV. *A diameter perpendicular to a chord bisects the chord and the arcs intercepted by it.*

In Fig. 93 let AB be the chord intercepting arcs ADB and ACB in $\odot O$, and CD the diameter perpendicular to chord AB at P. To prove that CD bisects AB and the two arcs we must show that $\overline{AP} = \overline{BP}$, $\widehat{AD} = \widehat{BD}$, and $\widehat{AC} = \widehat{BC}$.

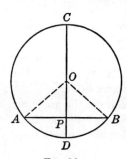

Fig. 93.

Draw the radii OA, OB. Then, since $OP \perp AB$, triangles OPA, OPB are right triangles. And since the radii $OA = OB$, and of course $OP = OP$, the hypotenuse and a leg of one equal those of the other. The right triangles are therefore congruent (XXI, 41) and hence the corresponding sides $\overline{AP} = \overline{BP}$, as was to be proved.

Also in the two congruent triangles, the corresponding acute $\angle AOP = \angle BOP$, that is, the central $\angle AOD = \angle BOD$, and hence, by II above, $\widehat{AD} = \widehat{BD}$, as was to be proved. Also since $\angle AOP = \angle BOP$, their supplements $\angle AOC = \angle BOC$, and hence, again by II, the intercepted $\widehat{AC} = \widehat{BC}$, as was to be proved. Thus the entire theorem is proved.

Now also, since $\widehat{AC} = \widehat{BC}$ and $\widehat{AD} = \widehat{BD}$, then by the equality addition axiom, the sum $\widehat{CAD} = \widehat{CBD}$. That is,

V. *A diameter bisects the circle*,

forming the *semi-circles* described in article 53.

VI. *In the same circle, or equal circles, equal chords are equidistant from the center.*

In Fig. 94 let chord $AB = CD$; then we have to prove that the perpendicular distance $OE = OF$.

Draw the radii OA, OC, forming the right triangles OEA, OFC (since by the definition following XI, 29, the *distance* from a point to a line is the *perpendicular*). Then according to V above, AB, CD are bisected at E, F, and as given, $AB = CD$, hence $AE = CF$ as halves of equals. Also the radii $OA = OC$. Hence the right triangles OEA, OFC have a leg and hypotenuse equal each to each and are congruent. Therefore the other legs are equal, $OE = OF$, as was to be proved.

From this result it is seen at once that also *chords equidistant from the center are equal; unequal chords are not equidistant from the center; and the nearer the center the greater is the chord*. Therefore, since a diameter is a chord at the center, *a diameter of a circle is greater than any other chord*.

We now demonstrate a very interesting theorem, one of the most important in the theory of the circle. It is

VII. *Through three points not in the same straight line one and only one circle can be drawn.*

In Fig. 95 let A, B, C be the three points not in the same straight line. We are to prove that a circle can be drawn to pass through A, B, and C, and that this is the only one that can be drawn.

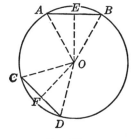
Fig. 94.

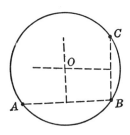
Fig. 95.

Draw the lines AB and BC, and at the middle point of each erect a perpendicular. Then since the segments AB, BC do not form a single straight line, the perpendiculars are not parallel, and hence will meet at some point, say O.

Then the point O is in the perpendicular bisector of AB and hence is equidistant from A and B, and similarly O is equidistant from B and C (IX, 29).

Therefore, O is equidistant from A, B and C, and hence, by the definition of a circle, a circle can be drawn with center O and radius $OA = OB = OC$ passing through A, B and C, as was to be proved.

Now the center of this circle must be in the perpendicular bisectors of both chords *AB*, *BC* (by IV above), and hence must be in their intersection *O*. But as the two perpendiculars can intersect in only one point (II, 22), there can be only one center for a circle to pass through *A*, *B* and *C* and hence only one such circle, as was also to be proved.

From this result we have at once, also,

VIII. *Two circles can intersect in only two points.*

For, if any two circles have *three* points in common (intersections), then by VII they must be the same circle.

This result answers the question which arose in connection with I above, and completes a trio of properties stated by II, 22; I, 54; and now VIII, 54.

Fig. 96(*a*) shows the intersection of two circles, at the points *A*, *B*.

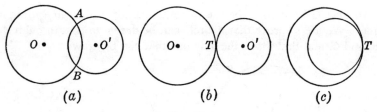

Fig. 96.

If the centers *O*, *O'* of the two circles in Fig. 96(*a*) be separated farther and farther, the intersection points become closer and closer together, until finally they coincide at the single point *T*, as in (*b*). The two circles then just touch, in one point, and are said to be *tangent* to each other.

Since in (*b*) each circle is outside the other, the circles are said to be *tangent externally*.

If ⊙*O'* is pushed farther and farther into ⊙*O* in (*a*) they finally have the positions shown in (*c*) where they again touch at one point (*T*). Since one circle is here inside the other they are said to be *tangent internally* at the point *T*.

We next investigate some of the properties of tangent lines and tangent circles, beginning with the fundamental theorem.

IX. *A straight line perpendicular to a radius at its extremity is tangent to the circle.*

In Fig. 97 let AB be the straight line perpendicular to the radius OT at its extremity T. To prove that AB is tangent to the circle we must show that it touches the circle at only one point.

Draw *any other line OD* from O to AB, meeting the circle at C. Then according to XI, 29, $OD > OT$ the perpendicular. That is, $OD > OC$ and hence D is outside the circle. Hence any, and every, such point as D on AB is outside the circle. But T, the end of the radius, is, by definition of radius, on the circle. Therefore, AB touches the circle at only one point T, and hence AB is tangent to $\odot O$ at T, as was to be proved.

We have also, conversely,

X. *A tangent to a circle is perpendicular to the radius drawn to the point of contact.*

XI. *The tangents from an external point to a circle are equal, and make equal angles with the line joining the point to the center.*

In Fig. 98 AB and AC are the tangents from the external point A to $\odot O$, and AO the line to the center. We have to prove that $AB = AC$ and $\angle BAO = \angle CAO$.

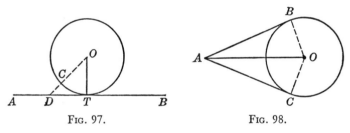

Fig. 97.   Fig. 98.

Draw the radii OB, OC, which are equal. Then $OB \perp AB$ and $OC \perp AC$, according to X above, and hence $\triangle ABO$, ACO are right triangles. Also the radii OB, OC are equal, and of course $AO = AO$. Hence the two right triangles have a leg and hypotenuse equal and are congruent (XXI, 41), and hence the corresponding parts $AB = AC$ and $\angle BAO = \angle CAO$, as was to be proved.

XII. *The line of centers of two intersecting circles is the perpendicular bisector of their common chord.*

In Fig. 99 let C, C' be the centers of the circles intersecting at A, B and each having AB as a (common) chord. We then have to prove that CC' is the perpendicular bisector of AB.

Draw the radii CA, CB of $\odot C$ and $C'A$, $C'B$ of $\odot C'$. Then $CA = CB$ and $C'A = C'B$, and hence the points C, C' are each equidistant from the ends of AB. According to IX, 29, therefore, CC' is the perpendicular bisector of AB, as was to be proved.

If two circles are tangent, as in Fig. 96(b), a line tangent to $\odot O$ at T is also tangent to $\odot O'$ at T, since T is the only common point of the two circles and a line touching one at T only, also touches the other at T only. By definition this line is tangent to each circle; it is therefore the *common tangent* to the two circles. Thus in Fig. 100 the line AB is the common tangent to the circles C, C' at the point T.

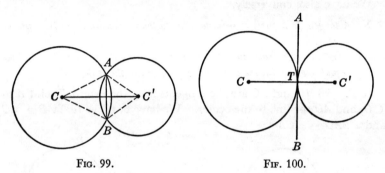

Fig. 99.   Fif. 100.

XIII. *The line of centers of two tangent circles passes through their point of contact.*

Since AB is the common tangent of the two circles C, C', a line perpendicular to AB at T will pass through the center of each circle, according to IX, X above.

But CC' also passes through these centers, and hence has two points in common with the perpendicular at T. It therefore coincides with the perpendicular (I, 22). Hence CC' passes through T, as was to be proved.

XIV. *Parallel lines intercept equal arcs on a circle.*

In Fig. 101 let the tangents AB, CD and the secants EF, GH be parallels. We have to prove that the pairs of arcs JE, JF; JG, JH; EG, FH; EK, FK; GK, HK; JEK, JFK are equal two by two.

Draw $JK \perp AB$; then also $JK \perp CD$ (XVII, 31), and being perpendicular to the tangents at the points of tangency, is a diameter, by IX, X above. Hence $\widehat{JEK} = \widehat{JFK}$, by V.

Also, according to IV, JK bisects EF and $\overset{\frown}{EJF}$, or, $\overset{\frown}{JE} = \overset{\frown}{JF}$; and similarly $\overset{\frown}{GK} = \overset{\frown}{HK}$.

Subtracting the equals $\overset{\frown}{JE}$, $\overset{\frown}{JF}$ from the equals $\overset{\frown}{JEK}$, $\overset{\frown}{JFK}$ we have the remainders $\overset{\frown}{EK} = \overset{\frown}{FK}$; and similarly $\overset{\frown}{JG} = \overset{\frown}{JH}$ (by the equality subtraction axiom).

Again, subtracting the equals $\overset{\frown}{JE}$, $\overset{\frown}{JF}$ from the equals $\overset{\frown}{JG}$, $\overset{\frown}{JH}$ we have also $\overset{\frown}{EG} = \overset{\frown}{FH}$.

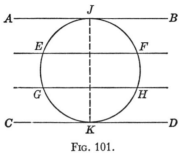

Fig. 101.

Thus the arcs intercepted between parallel tangents, parallel secants, or a parallel tangent and secant, are equal, as was to be proved.

It is to be noted that this theorem is similar to XXXIV, 46.

55. Measurement and Limits. In general usage the verb "*to measure*" means "to find the size or dimensions of," as applied to common objects, distances, etc. Thus to measure a rod or line is to find its length; to measure an angle is to find its size; etc. But when the operation of measuring is in either case carefully considered it is seen to consist simply in finding the number of times some previously chosen standard of length, angle, etc., is contained in the length, angle, etc., to be measured. Thus we lay a foot rule, yard or meter stick, along the line and count off the number of *inches* or *feet* or the number of lengths of the rule or stick, contained in the given line; or we lay down a protractor on the figure of the angle with its center at the vertex and count the *degrees* on the scale contained between the sides of the angle.

Thus in each case the *inch, foot, degree,* etc., must be known *in advance*. But each of these is itself simply a length or angle. Therefore measurement consists simply in finding the *number* of times a chosen standard of length, angle, etc., is contained in the length, angle, etc., to be measured.

This standard is called the *unit of measure* and the number is called the *numerical measure*, or simply the *measure*, of the thing measured. Thus the units of measure of length are the centimeter, inch, foot, yard, meter, etc. The units of measure of angle are the second, minute, degree, quadrant and circle.

Anything which can be measured is called a *quantity*. Thus length, area, volume, weight and the like are quantities.

The word "quantity" is also used in the sense of expressing "how much," but this is what we have defined above as the numerical *measure* of a thing. In mathematics and in science generally, "quantity" is used to designate the thing or entity itself, and the idea of "how much" is expressed by the term "measure."

If two quantities to be measured, say two lengths, each contain the same unit of measure a whole number of times without a remainder, that is, the numerical measure of each is a whole number, the two are said to be *commensurable*. The unit of measure is called their *common measure* and each quantity is said to be a *multiple* of this measure.

It may be that with a certain unit of measure one or the other, or both, of the two quantities may not be a multiple of the unit. Thus one length may be $2\frac{1}{3}$ yards and the other 4 yards. In this case the *yard* is not their common measure. But $2\frac{1}{3}$ yards = 7 feet, and 4 yards = 12 feet. Thus if the foot is the unit, the measure of one of the two lengths is 7 and of the other 12, both whole numbers. Again one angle may be 2° 12′ and another 50′ 36″. Here neither quantity is a multiple of the degree or of the minute. If the second is taken as the unit, however, the measure of one angle is 2° 12′ = 7920″ and of the other 50′ 36″ = 3036″. Thus each quantity is a multiple of the second, and the second is their common measure.

If two quantities have *no* common measure at all they are said to be *incommensurable*.

If a fraction is formed by writing the numerical measure of one quantity as numerator and that of another quantity *of the same kind* in terms of the same unit as denominator, the fraction is called the *ratio* of the two quantities, or more specifically, the ratio of the first to the second. Similarly, we can form the ratio of the second to the first, or the ratio of any quantity to another, provided we know the numerical measure of both.

Thus the ratio of the $2\frac{1}{3}$ yd. = 7 ft. to the 4 yd. = 12 ft. length is $\frac{7}{12}$; the ratio of the 50′ 36″ = 3036″ angle to the 2° 12′ = 7920″ angle is $\frac{3036}{7920} = \frac{253}{660}$.

Since the numerical measure of each of two quantities in terms of the *same unit* must be known in order to find their ratio, the ratio of two incommensurable quantities *cannot be found*. We shall find later that the diameter and circumference of a circle, for example, are incommensurable.

By taking the unit of measure *sufficiently small*, however, as in the example given above, the ratio of two incommensurable quantities can be expressed as *nearly* to the true value as may be desired. Thus an approximate value of the ratio can be found which shall differ from the true value by an amount less than any specified amount, however small.

When one quantity or number can be made to take on different values so as ultimately to differ from any fixed constant value by an amount less than any specified amount, however small, the fixed value or quantity is called the *limit* or *limiting value* of the variable quantity. The variable (changing) quantity is then said to *approach* the constant value *as its limit*.

Thus the true value of the ratio of two incommensurable quantities is the limit of the varying sequence of values found by taking their unit of measure smaller and smaller.

Another example, which we shall have occasion to study later, is the case of a polygon inscribed in a circle. Suppose that in Fig. 92(c) the number of sides of the inscribed polygon is increased; its perimeter or boundary will become closer to the circle. If the number of sides is indefinitely increased, the vertices being always on the circle, the polygon approaches closer and closer to the circle and it can ultimately be made to differ from the circle by less than any specified amount, however small. The polygon then approaches the circle as a limit; the circumference of the circle is the limit of the perimeter of the polygon.

56. Measure of Angles. By the term "measure of angles" it should be understood that we mean here not only the statement of the size of an angle in degrees, minutes, seconds, but also the method of determining the size of angles and the study of the relations between central and inscribed angles of circles and the various arcs and lines associated with them.

We begin the study of this important subject with the theorem which states that

XV. *In the same or equal circles two central angles have the same ratio as their intercepted arcs.*

In Fig. 102 let the circles, C, C' be equal; in (b), (a) let $\angle ACB$, $\angle A'C'B'$ be two central angles whose arcs AB, $A'B'$ are commen-

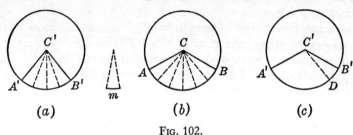

Fig. 102.

surable; and in (b), (c) let $\angle ACB$, $\angle A'C'B'$ be two central angles whose arcs $\widehat{AB}$, $\widehat{A'B'}$ are incommensurable. We have to prove in each case that the ratio $\dfrac{\angle A'C'B'}{\angle ACB} = \dfrac{\widehat{A'B'}}{\widehat{AB}}.$

Consider first the commensurable arcs in (a) and (b). Suppose the arc of length m to be the common measure of $\widehat{AB}$ and $\widehat{A'B'}$; then m is contained in each arc a whole number of times (by definition of common measure), suppose, for example, 4 times in $\widehat{A'B'}$ and 6 times in $\widehat{AB}$. Then the ratio of the arcs is $\dfrac{\widehat{A'B'}}{\widehat{AB}} = \dfrac{4}{6} = \dfrac{2}{3}$ (by definition of ratio).

Lay off the arc m on each arc and at the division points draw radii, dividing the central angles into the same number of parts as the arcs. Then these smaller central angles are equal (II, 54). Then the ratio of the central angles is $\dfrac{\angle A'B'C'}{\angle ACB} = \dfrac{4}{6} = \dfrac{2}{3}.$

But the ratio of the arcs is also $\frac{2}{3}$. Therefore, by the equality axiom,

$$\dfrac{\angle A'C'B'}{\angle ACB} = \dfrac{\widehat{A'B'}}{\widehat{AB}} \qquad (1)$$

—Q.E.D.

If the arcs are incommensurable then they will have no common measure (article 55) and cannot be divided into equal parts as was done above, but still some part $\overarc{A'D}$, in (c), can be found which is commensurable with $\overarc{AB}$ and then, as proved above, formula (1),

$$\frac{\angle A'C'D}{\angle ACB} = \frac{\overarc{A'D}}{\overarc{AB}} \tag{2}$$

If, however, the unit of measure be taken smaller and smaller, then the part left over, $\overarc{DB'}$, also becomes smaller and smaller; and $\overarc{A'D}$ approaches $\overarc{A'B'}$, and $\angle A'C'D$ approaches $\angle A'C'B'$, as limits. Ultimately, therefore, the ratio $\dfrac{\angle A'C'D}{\angle ACB}$ can be made to differ as little as we please from the ratio $\dfrac{\angle A'C'B'}{\angle ACB}$, and likewise for the ratios $\dfrac{\overarc{A'D}}{\overarc{AB}}$ and $\dfrac{\overarc{A'B'}}{\overarc{AB}}$ (article 55). In the limit, therefore, according to formula (2) above,

$$\frac{\angle A'C'B'}{\angle ACB} = \frac{\overarc{A'B'}}{\overarc{AB}}. \qquad \text{—Q.E.D.}$$

This demonstration is somewhat long and involved, but the results are very important.

The fact stated by the theorem in the form of the last proportion above can be stated in another form by subjecting the proportion to one of the transformations studied in algebra. Thus according to algebra the last proportion can be written as

$$\frac{\angle A'C'B'}{\overarc{A'B'}} = \frac{\angle ACB}{\overarc{AB}}.$$

This states, *if the proper unit of measure is used for arc and angle*, that the ratio of any central angle to its intercepted arc is the same as for any other central angle and its arc, that is, the *same for all central angles*.

If the unit of central angle measure is taken as *one degree* (see article 27) then the *unit of arc* measure is the arc intercepted by the central angle of one degree. This unit of arc is called the *arc degree*. Hence the *numerical measure* of a central angle expressed in degrees is equal to the numerical measure of its intercepted arc expressed in arc degrees. This result can be expressed by saying that

XVI. *A central angle is measured by its intercepted arc.*

That is, a central angle and its intercepted arc contain the same number of degrees, minutes or seconds.

Therefore, a right central angle intercepts an arc (quadrant) of 90 arc degrees; a semi-circle contains 180° of arc; a circle contains 360° of arc.

Expressed differently, this amounts to what has already been said in article 27. We have here, however, a sure and logical basis for what was there simply stated as a rule, with no reason given.

The result XVI just found for *central* angles will now be used to show how *inscribed* angles (Fig. 92(c)) are measured.

XVII. *An inscribed angle is measured by half its intercepted arc.*

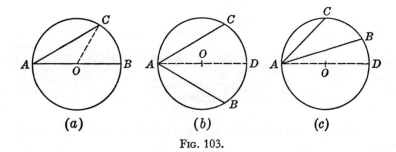

Fig. 103.

There are three ways in which an inscribed angle may be drawn. These are shown in Fig. 103. Thus

(a) the center (O) is on one side of the angle;
(b) the center is between the two sides of the angle;
(c) the center is outside the angle.

In each case let $\angle BAC$ be the inscribed angle with vertex A (on the circle) and intercepted arc $\overarc{BC}$. We then have to prove that $\angle A$ is measured by $\frac{1}{2}\overarc{BC}$.

In (a) draw the radius OC, and in (b) and (c) draw the diameter AD, through the center O.

Then in (a) $OA = OC$ and $\triangle AOC$ is isosceles, and hence $\angle C = \angle A$ (VII, 39). Also exterior $\angle BOC = \angle A + \angle C$ (III, 39). That is, $\angle BOC = 2\angle A$ and hence $\angle A = \frac{1}{2}\angle BOC$.

But central $\angle BOC$ is measured by its arc $\overarc{BC}$; therefore $\angle A$ is measured by $\tfrac{1}{2}\widehat{BC}$, as was to be proved. This means that $\angle A$ contains half as many degrees as $\widehat{BC}$.

In (b), according to the result just proved, $\angle DAC$ is measured by $\tfrac{1}{2}DC$, and similarly $\angle BAD$ by $\tfrac{1}{2}BD$. Hence, adding, the sum $\angle BAD + \angle DAC = \angle A$ is measured by the sum $\tfrac{1}{2}\widehat{BD} + \tfrac{1}{2}\widehat{DC} = \tfrac{1}{2}(\widehat{BD} + \widehat{DC}) = \tfrac{1}{2}\widehat{BC}$, as was to be proved.

Similarly in (c) $\angle DAC$ is measured by $\tfrac{1}{2}\widehat{DC}$, $\angle BAD$ by $\tfrac{1}{2}\widehat{BD}$, and the difference $\angle A$ is measured by $\tfrac{1}{2}\widehat{BC}$. Thus the theorem is proved for any inscribed angle of a circle.

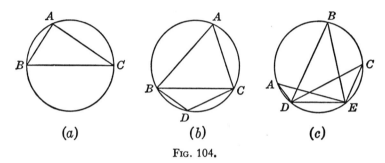

FIG. 104.

Consider now Fig. 104 in which

(a) $\angle A$ is inscribed in the semi-circle BAC;

(b) $\angle A$ is inscribed in the arc $\widehat{BAC}$ greater than a semi-circle and $\angle D$ in $\widehat{BDC}$ less than a semi-circle;

(c) $\angle A$, $\angle B$, $\angle C$ are inscribed in the same $\widehat{DE}$.

We have then, according to XVII:

XVIII. *Any angle inscribed in a semi-circle is a right angle.*

For it is measured by half a semi-circle, and a semi-circle contains 180 degrees.

XIX. *An angle inscribed in a segment of a circle greater than a semi-circle is acute, and one inscribed in a segment less than a semi-circle is obtuse.*

For the first is measured by less than half a semi-circle, as $\angle A$ in (b); and vice versa for the second, as $\angle D$ in (b).

XX. *Angles inscribed in the same or equal segments are equal.*

For each is measured by half the same arc, as ∠s A, B, C in (c). The result XVIII is of very great importance.

XXI. *An angle formed by a tangent and a chord from the point of tangency is measured by half the arc intercepted by the chord.*

In Fig. 105 let AB be the tangent at T and TC the chord from T. We then have to prove that $\angle BTC$ is measured by $\frac{1}{2}\widehat{TC}$.

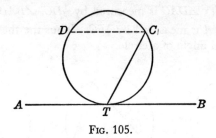

Fig. 105.

Draw the chord $CD \parallel AB$. Then the alternate-interior $\angle BTC = \angle DCT$; and also $\widehat{TC} = \widehat{TD}$ (XIV, 54). But $\angle DCT$ is measured by $\frac{1}{2}\widehat{DT}$, by XVII above. Hence the equal $\angle BTC$ is measured by $\frac{1}{2}\widehat{TC}$, the equal arc, as was to be proved.

57. Circle Constructions. To construct any small circle it is of course only necessary to set the compass at the proper radius and draw the circle in the manner already familiar, while larger circles are constructed by using larger compasses, or by holding fixed at the center, one end of a stretched string or cord whose length is the radius, and sweeping the other end around with the pencil or marker attached to it.

It may be required, however, to draw a circle which shall satisfy certain other specified conditions, such as passing through certain specified points or fitting into a certain position in some drawing already partly completed, etc. We now give the solutions of several problems of this kind, showing something of the general methods of procedure in such cases.

XXII. *To construct a circle which has a given radius and passes through two given points.*

In Fig. 106(a) let A, B be the given points and CD the given radius. Draw the line AB, and erect its perpendicular bisector (XIII, 30).

ART. 57   SOME PROPERTIES OF THE CIRCLE   121

With center at A and radius equal to CD draw an arc intersecting the perpendicular bisector at O. With center O and radius $OA(=CD)$ describe the circle through A and B, the required circle.

Proof. By construction $OA = CD$, and by IX, 29, $OB = OA$. Hence, $\odot O$ through A also passes through B, as required.

XXIII. *To construct a circle which shall pass through three given points not in the same straight line.*

In Fig. 106(b) let A, B, C be the three points. Draw AB and BC and erect their perpendicular bisectors, meeting at some point O. With O as center and radius OA describe a circle; it is the required circle through A, B and C.

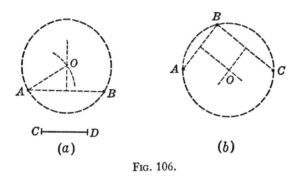

Fig. 106.

Proof. Since O is in the perpendicular bisector of AB, then $OA = OB$, and similarly $OB = OC$ (IX, 29). Hence the $\odot O$ of radius OA passes also through B and C.

This construction is based upon theorem VII, 54. The condition stated in that theorem is sometimes expressed by saying that *three points determine a circle.*

The problems XXII, XXIII above are a companion pair. It is to be noted that three things are given in each: three points in the second, and in the first two points and the radius. This is a general property of the circle, which is expressed by stating that *three conditions are required to describe a circle*, that is, to give its position and size.

We give next the constructions of two circles associated with a triangle, and later certain triangles associated with a circle.

XXIV. *To inscribe a circle in a given triangle.*

In Fig. 107 let △ABC be the given triangle, and draw the bisectors of ∡A, B, meeting at O (by XXVI, 33). From O draw OF ⊥ AB (by XII, 30). With O as center and OF as radius describe circle O(DEF). This is the required circle.

Proof. The point O is equally distant from the sides of the triangle (XIII, 40); that is, OD = OE = OF. Hence the circle with OF as radius touches each of the three sides, and is therefore, by definition, the inscribed circle.

XXV. *To circumscribe a circle about a given triangle.*

In Fig. 108 let △ABC be the given triangle. Consider the three vertices of this triangle as three given points through which a circle is to be drawn, and proceed as in XXIII above.

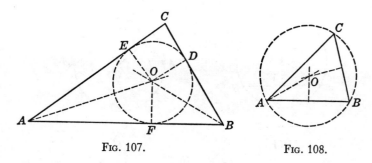

FIG. 107.   FIG. 108.

The last two problems above form a companion pair. The center O of the inscribed circle in Fig. 107 is called the *in-center* of △ABC, and the center O of the circumscribed circle in Fig. 108 is called the *circum-center* of △ABC.

We give next an important circle construction closely related to problem XIII, 30, which was there postponed until the circle should have been studied. It is:

XXVI. *To erect a perpendicular at the end of a given line segment.*

In Fig. 109 let AB be the given segment and let it be required to erect the perpendicular at B, for example.

Construction. Take any convenient point C not on AB, and with C as center and the distance CB as radius describe an arc or circle through B intersecting AB at D. Draw DC and extend it to meet the arc at E, and then draw BE. BE is the required perpendicular.

Proof. Since C is the center DE is a diameter of the construction

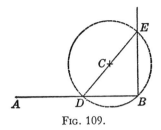

Fig. 109.

circle, and hence the arc DBE is a semi-circle. According to XVIII, 56, therefore, $\angle DBE$ is a right angle, and hence, by definition, $BE \perp AB$.

The next three construction problems form a related group on the construction of tangents to circles.

XXVII. *To draw a tangent to a circle at a given point on the circle.*

In Fig. 110 let T be the given point on $\odot O$. Then according to theorem X, 54, it is only necessary to draw $AB \perp OT$ at the end T, and this is done at once as in XXVI above.

XXVIII. *To draw a tangent to a circle from a given external point.*

In Fig. 111 let $\odot C$ be the circle and P the external point. Draw CP and on CP as diameter describe a circle, cutting $\odot C$ at A and B. Draw CA and PA; then PA is the required tangent. Similarly another tangent PB may be drawn from P.

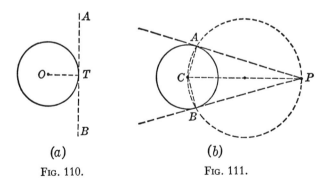

(a) (b)

Fig. 110. Fig. 111.

For, $\angle PAC$ is inscribed in the semi-circle PAC, and hence, according to XVIII, 56, $PA \perp CA$. And since CA is a radius, therefore PA is a tangent (X, 54).

It may appear at first sight that in order to draw a tangent to a circle it is only necessary to lay a straight-edge in such a position that it appears as just touching the circle, and then to draw a line by the straight-edge. If this is tried, however, it will be seen that it is very difficult (theoretically impossible) to tell when the straight-edge and the circle *just touch, at one point*. But it is easy to draw a line through a fixed point or to draw a perpendicular to a line at a given point. And since a tangent is perpendicular to a radius where it meets the circle it is only necessary to fix the point and draw the radius, and then draw the perpendicular to the radius at that point, which is the tangent. This is the procedure in both the last two cases.

A line which is tangent to two or more circles is called their *common tangent*. This definition has already been given for a particular case (see XIII, 54), but the two circles need not themselves be tangent to one another. Thus in Fig. 112, *AB* and *CD* are both tangent to both circles *O*, *O'*; and so also are *EF* and *GH*.

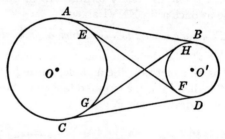

Fig. 112.

The pair of tangents *AB*, *CD* are called the *common external tangents* and the pair *EF*, *GH* the *common internal tangents* of the two circles *O*, *O'*.

If the circles *O*, *O'* are thought of as two wheels the common external tangents will represent a stretched *open belt* and the common internal tangents a *crossed belt*.

Having seen the requirements for drawing a single tangent to one circle we are now prepared to appreciate the problem a mechanical draftsman has to solve when he is required exactly

XXIX. *To locate the center of a given circle.*

This is easily done. Thus let the circle of Fig. 113 be the given circle, with its center not known. Mark any three points on it, as *A*,

B, C; join adjacent points by chords AB, BC and erect the perpendicular bisectors of the chords. Their intersection O is the center of the circle.

For, according to IX, 29, a point in the perpendicular bisector of a line is equidistant from its ends. Hence O is equidistant from A, B, C and is therefore the center of the circle on which these three points lie.

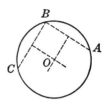

Fig. 113.

This is a construction which is often useful.

58. Illustrations and Applications. (i) *The Protractor.* This instrument has already been described, in article 27. It is now seen on the basis of XVI, 56, that the circular edge of the instrument is divided into 180 *arc* degrees to correspond with the (semicircular) straight angle of 180 *angle* degrees, and similarly in the case of any other angle whose vertex is at the center of the protractor circle.

(ii) *Angle of Elevation.* An object or a line whose direction from an observer makes an angle with the horizontal (or the vertical, article 24) is said to be *elevated* above the horizon and the angle between the direction and the horizontal is called the *angle of elevation.*

Thus if an observer is at O, Fig. 114, the sun is seen at sunrise on the horizon in the direction OH. At some time in the forenoon it is seen in some such direction as OS, and its *elevation* at that time is the angle HOS. If an object is seen directly overhead (on the vertical) its elevation is 90°. On the horizon the elevation is, of course, zero.

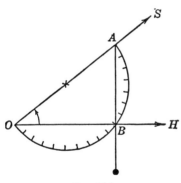

Fig. 114.

The horizontal and vertical are perpendicular and are determined as shown in article 43, Ex. 11. Any other angle of elevation is easily determined as follows, see Fig. 114.

Suspend a small weight by a thread of convenient length from one end of the straight edge of a protractor, as at A, and hold the protractor so that the string hangs freely and just touches the flat face of the protractor. Sight along the edge OA at the object whose elevation is desired, and note the point B where the string crosses the edge of the scale. Read the arc AB on the scale in degrees.

The direction AB is then the vertical (plumb-line), OB is the horizontal and is perpendicular to AB ($\angle B = 90°$ by XVIII, 56), and the elevation $\angle HOS$ is *half* the scale reading $\overset{\frown}{AB}$. For, $\angle BOA$ is an *inscribed* angle (vertex on circle at O) and is therefore measured by half its intercepted arc AB (XVII, 56).

(iii) *Latitude and Longitude.* It is assumed that the reader is familiar with the spherical (globular) shape of the earth and knows something of latitude and longitude as explained in geography or arithmetic (see the ARITHMETIC of this series, Chapter XV), and these will be simply defined here without detailed explanation, for reference in the following application.

The spherical earth rotates once a day about a line through its center (diameter), called the *axis* of rotation, and its ends, at the surface, are called the *poles* of the earth. The pole which we face when the right hand is turned in the direction of rotation (the sunrise) is called the *north* pole and the other the *south* pole.

A circle drawn (or imagined) around the surface of the earth midway between the poles is called the *equator*, and other circles parallel to the equator between the equator and either pole are called *parallels of latitude* and run in the east-west direction.

A certain point on the equator is agreed upon as a standard reference point, and the equator (circle) is divided into degrees, beginning at that point. These divisions are said to measure *longitude* from the reference point and circles on the earth passing through these divisions and through both the poles are called *meridians of longitude*. The meridians of longitude thus run in the north-south direction.

Any meridian circle (of longitude) is cut by all the parallels of latitude on each side (north and south) of the equator. The parallels thus divide the meridian circles, and these meridian divisions are marked in degrees and read from the equator toward each pole, 90° between the equator and either pole. The *latitude* of a place on the earth is thus measured in degrees (or ° ′ ″) *north* (N) or *south* (S) from the equator.

Similarly, the meridian circles divide the equator and all the parallels, and longitude is measured in degrees *east* (E) or *west* (W) from the chosen reference point on the equator. This point is by international agreement taken as the point where the meridian line of the British National Observatory, at Greenwich, England, crosses the equator. Longitude is then measured up to 180° in each direction around to the opposite side of the earth, where the same meridian again cuts the equator.

The position of any place on the earth is stated or shown on a map by giving *both* its latitude (N or S) and its longitude (E or W). Thus the latitude of the government reference marker for New York City is 40° 40′ 47.17″ N and its longitude is 73° 58′ 41.00″ W.

(iv) *Determination of Latitude.* The latitude of a place on the earth is found by a geometrical measurement of angle, while its longitude is found from its time and the relation between time and longitude as expressed in arithmetic (see the reference above). We shall consider here the determination of latitude only.

The axis of the earth points toward two certain points in the heavens, and at the place toward which the north end of the axis (north pole) points is a fixed star called *Polaris* or the *North Star*. (There is no such south polar star.) If an observer is at the north pole the North Star is directly overhead (elevation 90°) and if he is at the equator the Star is on the horizon (elevation 0°). At intermediate places, depending on the latitude, the elevation of the Star may have any value from 0 to 90°. In fact, *the latitude of a place is equal to the elevation of the pole star,* as we shall now show.

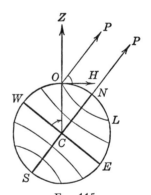

Fig. 115.

In Fig. 115 let the circle represent the earth, the line *NS* its axis, pointing toward the North Star, and *WE* its equator; the curved lines are then parallels of latitude.

If an observer is at *N* the pole star is to him directly overhead, in the direction *NP*. To an observer at *O*, however, the direction *OZ* is directly overhead (*the zenith*), and *OH* ⊥ *OZ* is the horizontal; the star is then in the direction *OP* ∥ *NP*, and ∠*HOP* is its elevation. The circle *NOW* is his meridian of longitude, *OL* is his parallel of latitude, and ∠*WCO* is his latitude, measured north from the equator.

Now since *OP* ∥ *CP* and *CW* ⊥ *CP* then also *CW* ⊥ *OP* (XVII, 31). Also the horizon *OH* ⊥ *CZ* the vertical. That is, the two sides of ∠*HOP* are perpendicular to the two sides of ∠*WCO*. Therefore, ∠*HOP* = ∠*WCO* (XXIII, 32). That is, *the elevation of the pole star* at any place *is equal to the latitude* of the place.

In order to determine the latitude of a place (north) therefore, it is only necessary to measure the elevation of the pole star, as in example (ii) above.

(v) *To Find a Circle from Its Arc.* Suppose a piece of the rim of a broken wheel is all that can be found of the wheel and it is desired to find the size of the wheel (circular) and locate its center. This is easily done as follows. In Fig. 116 let $\overset{\frown}{ABC}$ represent the available portion of the wheel. When found then the dotted circle of the figure will represent the complete circle of the wheel, O will be its center, OC its radius, and the diameter is twice the length of OC, or is at once found by measuring through the center O.

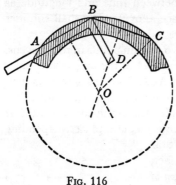

Fig. 116

Lay a carpenter's square (or draftsman's triangle) against the rim in the position ABD and measure the chord AB. Mark its middle point and from this point draw a line (shown dotted) parallel to BD. Similarly locate BC and its perpendicular bisector. The two perpendicular bisectors meet at the center O, and the line OC drawn to one of the original points C is the radius. This measurement is based upon proposition XXIX, 57.

(vi) *To Test a Semi-circular Groove.* In Fig. 117 let ABC be a groove in a block or beam (end view shown), supposed to be a semi-circle. To test it place a carpenter's square or draftsman's triangle in the groove as shown and move it around in all positions keeping the square directly across the groove. If it is truly a semi-circle the square will always touch at A, B and C (according to XVIII, 56). The same test will serve for a bowl, cup or any hole or cavity which is supposed to be half of a sphere.

Figure 118 shows how the same principle may be used to draw a circle without a compass or any round guide. Place two pins or tacks in the drawing board at A and B, whose distance apart is the desired diameter. Press the legs of the triangle against the tacks and move the triangle about so that the point P, the vertex of the right angle, moves from A to B. A pencil or marker at P will then trace the semi-circle *APB*, and by turning the triangle over the other half of the circle can be drawn.

The following example shows how a large circle can be drawn when its center is inaccessible or is not known.

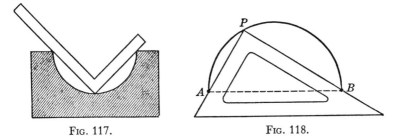

Fig. 117.   Fig. 118.

(vii) *To Lay Out a Large Circle without Locating the Center.* This operation is frequently necessary in building, surveying, landscape gardening, etc. In such cases the circle is generally too large to be described with a cord as in article 57, or the center is inaccessible because of some obstruction. In such a case, however, several points are known through which the circle must pass. The method is shown in Fig. 119.

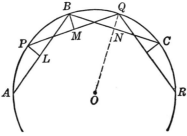

Fig. 119.

In Fig. 119 let *A*, *P*, *B* be three equidistant points through which the circle must pass. *P* is then the midpoint of $\widehat{AB}$ and when *PL* $\perp$ $\overline{AB}$, *L* is the midpoint of chord *AB* (IV, 54). So stretch a measuring tape between the known points *A* and *B* and draw *PL* $\perp$ $\overline{AB}$. Then swing the length *AL* about *P*, and *PL* about *B*, as centers, until they meet at *M*, and stretch the length *AB* along *PM* to *Q*. This fixes the point *Q* on the circle, with *M* as the midpoint of chord *PQ*.

In the same way, working from B, Q as centers and with the same radii, mark N and fix C. Continue, fixing R, etc., until enough points are located to draw the circle through them.

The original arc $\widehat{AB}$ and chord $\overline{AB}$ can be taken as short as desired and so the successive points may be located as near together as may be desired.

The chords AB, PQ, BC, QR, etc., are equal by construction, and so are the distances PL, BM, QN, etc. (each called the *rise* of its arc), or, what is the same thing, the remainders of each radius NO, etc., are all equal. That is, the chords are equal and equidistant from the center.

The construction is therefore based upon propositions IV, VI, 54.

(viii) *Eclipse of the Moon.* The explanation of a lunar eclipse affords an interesting use of the common tangents of two circles (or spheres). This is shown in Fig. 120. Here $\odot S$ represents the sun and $\odot E$ the

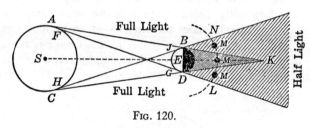

Fig. 120.

earth, and the dotted arc LMN, concentric with E, represents a part of the path of the moon M around the earth.

The sun, being many times larger than the earth, casts overlapping shadows, the rays of light past the earth's surface along the common tangents marking the edges of the shadows. The conical region (the earth being a sphere) BKD between the extended external tangents intersecting at K, is the region of darkness, full shadow, and is called the *umbra* of the earth's shadow. The region between the extended internal tangents and outside the umbra is the region of half light or semi-darkness and is called the *penumbra* of the earth's shadow. All space outside both internal and external tangents is in full sunlight.

As the moon in its circular path around the earth enters the penumbra and after it emerges from the umbra it is only partially illuminated by the sun and is said to be *partially eclipsed*. When it is in the umbra it is not illuminated at all, being in the complete darkness of night, and is said to be *totally eclipsed*. Before reaching L and after passing N it is of course illuminated by full sunlight and shines brilliantly, as a mirror, by reflected sunlight.

Observers on the illuminated part of the earth (opposite from the moon) cannot see the moon, of course, when it enters the shadows. A lunar eclipse is therefore visible only at night. It is, however, visible to every observer on the dark side of the earth at the same time.

The time, duration and other circumstances of a lunar eclipse are calculated by means of the known sizes, distances and motions of the sun, earth and moon and the angles made by the common tangents in Fig. 120, and can be completely predicted and described years in advance.

Fig. 120 is of course not drawn to scale. Since the distance SE is about 93 million miles and EM about a quarter of a million miles, while the diameter of the sun is about 900 thousand, that of the earth about 8 thousand and that of the moon a little over 2 thousand miles, the shadows are relatively long and narrow.

Remarks. The properties of the circle afford many uses and applications in art, architecture, engineering, etc., but the examples explained in detail in this article will be sufficient for our purposes. For others the reader is referred to books on the particular subjects of application in which he is interested.

The examples given in this article are based upon the *formal* properties of the circle (pertaining to forms and relations); examples illustrating the *numerical* properties will be given in the chapter on the *measurement* of the circle.

59. Exercises and Problems.

1. What is the angle between the hands of a clock when it is three o'clock? What is the angle at four o'clock; at half past eight; at 9:30?

2. What is the angular velocity of a wheel which rotates at 1000 R.P.M.? If the angular velocity of a shaft is 1800° per second, what is the speed in R.P.M.?

3. What are the angular velocities of the earth, and of the hour, minute and second hands of a watch?

4. If one wheel is running at 5 R.P.M. and another at 3 r.p.s., which is the faster?

5. Prove that the tangents at the ends of a diameter are parallel.

6. In the answers to the exercise of article 51, what appears to be the limit of the interior angle of a regular polygon as the number of sides is increased indefinitely?

7. What is the limit of the sum of the series $1 + \frac{1}{2} + \frac{1}{4} + \frac{1}{8} + \frac{1}{16} +$, etc., as the number of terms is increased indefinitely?

8. If an equilateral triangle is inscribed in a circle, how many degrees are there in each intercepted arc of the circle? Why?

9. A right angle with its vertex at the center of a circle intercepts 90° of arc at the circumference; how many if the vertex is on the circle?

10. If the diameter of a circle equals the hypotenuse of the draftsman's 30–60° triangle, and the triangle is placed on the circle with the long edge lying on the diameter, where is the vertex of the right angle? How many degrees of arc will each acute angle of the triangle intercept?

11. An arctic explorer got as near to the north pole as one-fortieth of the earth's circumference. How high above the horizon did the pole star appear to him?

12. At the northernmost tip of Scotland the elevation of the pole star is about 58° above the horizon. How far away is the north pole? (The earth's circumference is about 24,900 miles.)

13. Prove that if a polygon is inscribed in a circle the perpendicular bisectors of all its sides pass through one point.

Chapter 6

PROPORTION AND SIMILAR FIGURES

60. Ratio and Proportion. In article 55 the *ratio* of two numbers or two measured quantities of the same kind was defined as a fraction having one of the two numbers, or numerical measure of one of the measured quantities, as numerator and the other as denominator. Thus $\frac{3}{5}$ is the "ratio of 3 to 5"; $\frac{a}{b}$ is the ratio of a to b; $\frac{\overline{AB}}{\overline{CD}}$ is the ratio of the line segment AB ($\overline{AB}$ being the numerical measure of its length) to the segment CD; $\frac{\angle ABC}{\angle DEF}$ is the ratio of $\angle ABC$ to $\angle DEF$, when both are expressed in the same unit of measure. These may also be written as 3/5, 3:5, a/b; $a:b$, etc., the sign (/) or (:) having the same meaning as the ordinary fraction line (−).

In such a ratio as $\frac{a}{b}$ or $a:b$, the algebraic symbols a, b may represent any two numbers or the numerical measures of any two quantities of the same kind, as two lengths, angles, areas, volumes, etc.

In any ratio, as $\frac{a}{b}$ or $a:b$, the first number, as a, is called the *antecedent* and the other is called the *consequent*.

Suppose that the ratio of two certain numbers or quantities a and b, is the same as the ratio of two certain other numbers, c and d. This means that $\frac{a}{b} = \frac{c}{d}$. When *four* numbers are thus so related that the ratio of two of them is equal to that of the other two, the four numbers are said to be *in proportion* or *proportional*, and the equation which expresses this relation is called a *proportion*. Thus the numbers, a, b, c, d are in proportion, as expressed by the equation written above.

The proportion written above is also sometimes written in the form $a:b = c:d$ and the two forms are entirely equivalent. Such a proportion is expressed in words by saying that "*a* is to *b* as *c* is to *d*." This applies to either written form of the proportion.

In either form of the proportion the four numbers are said to be the *terms* or *members* of the proportion. The first and the last terms of a proportion, *a* and *d* above, are called the *extremes* and the two intermediate terms, *b* and *c* above, are called the *means*, of the proportion.

If the means of a proportion are equal, that is, both are the same number, as $x:y = y:z$, then this number, as y, is called the *mean proportional* between x and z. The number z is called the *third proportional* to x and y.

We have already had (article 56) a few examples of geometrical ratios and proportions; in this chapter we shall have many more. As many geometrical properties of figures are closely related by proportion it is important that the principles of proportion be familiar. The more immediately important of these are given in the next article.

61. Principles of Proportion. The principles of proportion are studied and fully explained in elementary algebra and will be simply stated here for later reference, without explanation or proof. (See the ALGEBRA of this series, Chapter 20.)

I. *In any proportion the product of the means equals the product of the extremes.*

Thus if $a:b = c:d$, then $ad = bc$.

II. *The mean proportional of two numbers is the square root of their product.*

Thus if $a:b = b:c$, then by I, $b^2 = ac$ and hence $b = \sqrt{ac}$.

III. *In any proportion the means or the extremes may be alternated, and both ratios may be inverted.*

This means that if $\frac{a}{b} = \frac{c}{d}$, then either $\frac{a}{c} = \frac{b}{d}$ or $\frac{d}{b} = \frac{c}{a}$; and also $\frac{b}{a} = \frac{d}{c}$.

IV. *If the product of any pair of numbers equals the product of any other pair, they form a proportion in which either pair may be the means and the other the extremes.*

Thus if $ad = bc$, then either $a:b = c:d$ or $b:a = d:c$. This is simply another way of stating I.

V. *The terms of a proportion are in proportion by composition.*

This means that if $\frac{a}{b} = \frac{c}{d}$, then $\frac{a+b}{b} = \frac{c+d}{d}$.

VI. *The terms of a proportion are in proportion by division.*

This means that if $\dfrac{a}{b} = \dfrac{c}{d}$, then $\dfrac{a-b}{b} = \dfrac{c-d}{d}$.

VII. *The terms of a proportion are in proportion by composition and division.*

This means that if $\dfrac{a}{b} = \dfrac{c}{d}$, then $\dfrac{a+b}{a-b} = \dfrac{c+d}{c-d}$.

VIII. *In a continued proportion the sum of the antecedents is to the sum of the consequents as any antecedent is to its consequent.*

This means that if $\dfrac{a}{b} = \dfrac{c}{d} = \dfrac{e}{f} = \dfrac{g}{h} =$ etc., then $\dfrac{a+c+e+g}{b+d+f+h} = \dfrac{a}{b}$, or $= \dfrac{e}{f}$, etc., and similarly for a continued proportion of any number of equal ratios.

62. Proportional Line Segments. We now consider the relations between certain line segments in a few geometrical figures.

IX. *A line drawn across two sides of a triangle parallel to the third side divides the first two sides proportionally.*

In Fig. 121 let $\triangle ABC$ be any triangle and DE a line parallel to the side AB. The theorem then states that $\overline{AD}:\overline{DC} = \overline{BE}:\overline{EC}$, that is, the two segments of one side have the same relation (ratio) as those of the other.

To prove this, suppose the segments AD, DC are commensurable and that CG is their common measure. Divide AC into segments of this length and suppose that there are, say, m parts in AD and n in DC. (In our figure $m = 2$, $n = 3$, but these may be any values, depending on the size of the triangle and the length of the common measure of

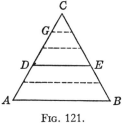

Fig. 121.

the segments.) Then since CG is the common measure of the lengths AD and DC, the numbers m, n are their numerical measures. The ratio of AD to DC is therefore equal to the fraction $\dfrac{m}{n}$. That is, $\dfrac{\overline{AD}}{\overline{DC}} = \dfrac{m}{n}$.

Now draw through the points of division of AC lines parallel to AB. They will divide BC into the same number of equal parts (XXXIV, 46) and in the same way as above we find that $\dfrac{\overline{BE}}{\overline{EC}} = \dfrac{m}{n}$.

We thus have the two ratios of segments both equal to the same fraction $\frac{m}{n}$. According to the equality axiom, therefore, they are equal to each other. That is, $\frac{\overline{AD}}{\overline{DC}} = \frac{\overline{BE}}{\overline{EC}}$, as was to be proved.

If the segments AD, DC are incommensurable, then we can proceed as in the proof of XV, 56, and show that the same proportion holds as above for the commensurable segments.

Now since $AD:DC = BE:EC$, then by composition (V, 61) we have also $\frac{AD + DC}{DC} = \frac{BE + EC}{EC}$. But $AD + DC = AC$ and $BE + EC = BC$. Therefore $\frac{AC}{DC} = \frac{BC}{EC}$. That is, we have also,

X. *If a line parallel to one side of a triangle cuts the other two sides the corresponding segments are proportional to the sides.*

The result just proved can be extended to obtain a relation somewhat similar to that expressed in XXXIV, 46. In that theorem the segments cut off on each transversal are *equal;* we can now show that when not equal on each transversal they are *proportional* on the two. Thus

XI. *If any number of parallels are cut by two transversals the corresponding segments on the two are proportional.*

In Fig. 122 let AB, CD be the transversals cutting the parallels AC, EF, GH, BD, etc., not equally spaced. We then have to show that $AE:CF = EG:FH$; $EG:FH = GB:HD$; etc.

Through A draw $AL \parallel CD$. Then since $AC \parallel JF \parallel KH$, etc., and also $AJ \parallel CF$, $JK \parallel FH$, etc., the quadrilaterals $ACFJ$, $JFHK$, etc.,

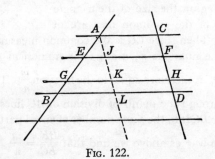

Fig. 122.

are parallelograms and hence as opposite sides $AJ = CF$, $JK = FH$, etc.

In the triangle ABL, according to IX, X above,

$$\frac{AE}{AJ} = \frac{EG}{JK}, \frac{AE}{AJ} = \frac{AG}{AK}, \frac{AG}{AK} = \frac{EG}{JK}, \text{etc.}$$

Hence

$$\frac{AE}{AJ} = \frac{EG}{JK} = \frac{GB}{KL}, \text{etc.,}$$

or

$$\frac{AE}{CF} = \frac{EG}{FH} = \frac{GB}{HD} = \text{etc.} \qquad \text{—Q.E.D.}$$

When three or more ratios are equal in this manner the members are said to be in *continued proportion*, as in VIII, 61.

63. Similar Figures. Two geometrical figures which have the same *shape*, regardless of their *sizes*, are said to be *similar*. Let us analyze this statement and see how to express similarity in terms of sides and angles.

Suppose a polygon of any number of sides, say five, to fix ideas, to be made of stiff steel wires, jointed at the vertices, and suppose that while lying on a smooth plane surface (as a table) some of the wires are pushed inward or pulled outward without being bent. The shape of the polygon will then be changed. The sides, however, remain straight and keep the same length, but they will turn about the joints at the vertices, and some or all of *the angles are changed*. Therefore, if the sides of a polygon are not changed, the angles must change in order to change the *shape*.

Next suppose the polygon to be rigid, that is, the sides stiff and the angles not jointed but, say, welded; and suppose that in some way, as by heating or cooling the sides all become longer or all become shorter without bending and without changing the angles. The polygon then becomes larger or smaller, that is, the *size* changes, but the shape remains the same. Suppose, however, one side lengthens (or shortens) more rapidly than the others; it is seen at once that the remaining sides are bent and the angles changed, so that the entire figure is distorted.

If, however, each side changes by the same fractional or relative amount, as, each side changes by one tenth or one half, of its length, then the shape does not change. Thus if the angles do not change, in order that the polygon maintain its same shape as the sides change, each side must be multiplied by the *same* amount; the new length divided by the old must give the same quotient for each side. That is, the *ratio* of the new length to the old must be the same for all the sides. This means that the sides of the one polygon are all *proportional* to the sides of the larger (or smaller) polygon when the size changes while the shape does not.

The results of the two sets of changes as analyzed above can now be summarized as follows: If the *shape* of a polygon is not to change the *angles must remain the same;* if the *size* (sides) changes without changing the shape the sides of the polygon of one size must be *proportional* to those of the other size.

We can now express our first definition of similar figures, given in the first paragraph of this article, in terms of sides and angles as follows:

Similar geometrical figures are those which have their corresponding ANGLES EQUAL *and their corresponding* SIDES PROPORTIONAL.

By *corresponding* angles or sides are meant those similarly situated, or in the same relative positions, in the two figures.

64. Properties of Similar Triangles. We have seen that for two figures to be similar two sets of conditions must be satisfied: *the angles must be equal* and *the sides must be proportional.* We now consider a theorem which states as hypothesis that the angles of a triangle are equal to those of another triangle, and concludes that the triangles are similar. In order to show that this relation follows, therefore, it is necessary to show that the sides are proportional. The theorem is

XII. *If the three angles of one triangle are equal to the three angles of another the triangles are similar.*

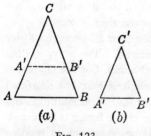

FIG. 123.

From the equality of the angles we thus have to prove the proportionality of the sides.

In Fig. 123 let $\triangle ABC$ and $\triangle A'B'C'$ be two triangles having $\angle A = \angle A'$, $\angle B = \angle B'$, $\angle C = \angle C'$. We then have to prove that $AB:A'B' = AC:A'C' = BC:B'C'$, this being a continued proportion between the three corresponding pairs of sides.

Suppose the triangle (*b*) placed upon the triangle (*a*) with $\angle C'$ on its equal $\angle C$ so that $C A'$ lies on CA and $C'B'$ lies on CB. Then in (*a*) the side $A'B'$ of (*b*) will be in the position shown by the dotted line.

Now in (a) $\angle A = \angle A'$ by hypothesis, and as these are corresponding angles formed by the transversal AC cutting the lines AB and $A'B'$, then $A'B' \parallel AB$ (by XX, 31). Therefore $AC:A'C = BC:B'C$ (by IX, 62), or, since $A'C$ is $A'C'$ and $B'C$ is $B'C'$, $AC:A'C' = BC:B'C'$.

By placing triangle (b) on (a) so that $\angle A'$ coincides with its equal $\angle A$, we find in the same way that $AB:A'B' = AC:A'C'$.

Combining the two proportions (equations) by the equality axiom we have $AB:A'B' = AC:A'C' = BC:B'C'$, as was to be proved.

Since now two triangles are shown to be similar if their angles are equal, and since if two angles of one triangle are equal to two angles of another the third must be equal because the sum is the same for both (180°), then we have also:

XIII. *If two angles of one triangle equal two angles of another the triangles are similar.*

And now, since in all right triangles one angle is always the same (90°), then by the last result we have also:

XIV. *If an acute angle of one right triangle equals an acute angle of another the right triangles are similar.*

We can now prove also:

XV. *If two triangles have an angle of one equal to an angle of the other and the including sides proportional the triangles are similar.*

This proposition should be compared with XII above. There the triangles were proved similar because their angles were equal; here we are to prove them similar because their sides are proportional. We must first show that in addition to the given equal angles, the other two are also equal each to each. For this purpose we again use Fig. 123, but with a different construction.

In Fig. 123 suppose this time that only $\angle C' = \angle C$, but that the including sides are proportional, that is, $AC:A'C' = BC:B'C'$.

Place the triangle (b) upon (a) so that the equal angles and including sides coincide and $A'B'$ is in the dotted position. Then since $AC:A'C = BC:B'C$, $A'B' \parallel AB$ (IX, 62). Therefore $\angle A' = \angle A$ and $\angle B' = \angle B$, as corresponding angles of parallels formed by a transversal.

But $\angle C' = \angle C$ by hypothesis; hence all the angles of (b) are equal to those of (a). According to XII above, therefore, $\triangle A'B'C'$ is similar to $\triangle ABC$, as was to be proved.

XVI. *If two triangles have their sides proportional they are similar.*

This proposition should also be compared with XII, the pair being exact companion or complementary theorems. There the triangles have all their angles equal each to each, and here all the sides are proportional. The two triangles are shown in Fig. 124.

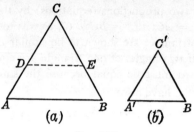

Fig. 124.

In Fig. 124 let (a) and (b) represent the triangles in which $AB:A'B' = AC:A'C' = BC:B'C'$. In (a) on CA lay off $CD = A'C'$, on CB lay off $CE = B'C'$, and draw DE. Then by hypothesis $AC:DC = BC:EC$ and of course in the two triangles in (a) the $\angle C$ is the same. Hence by XV $\triangle ABC$ is similar to $\triangle DEC$, and therefore by definition of similar figures, $AC:DC = AB:DE$, or since $DC = A'C'$ by construction, $AC:A'C' = AB:DE$.

But by hypothesis $AC:A'C' = AB:A'B'$; hence by the equality axiom (a proportion being an equation), $AB:A'B' = AB:DE$. In these last two ratios the antecedents are the same, and therefore the consequents are also equal, that is, $DE = A'B'$. Therefore in $\triangle DEC$ the three sides (two by construction) are equal to those of $\triangle A'B'C'$ and hence these triangles are congruent. But as shown, $\triangle ABC$ is similar to $\triangle DEC$; therefore $\triangle ABC$ is similar to $\triangle A'B'C'$, as was to be proved.

XVII. *If two triangles have their sides parallel or perpendicular each to each they are similar.*

For when the sides are parallel or perpendicular the angles are equal (XXII, XXIII, 32), and when the angles are equal the triangles are similar, by XII above.

The propositions XII, XIII, XV, XVI above state the chief conditions under which triangles are *similar*. These of course apply to right triangles as well as any others; they are, however, somewhat simplified in the case of right triangles, as illustrated by XIV.

ART. 64   PROPORTION AND SIMILAR FIGURES   141

These theorems should be compared with those of article 39 which state the conditions under which triangles are *congruent*. It is to be remembered that congruent triangles have their angles equal, and their sides equal instead of proportional.

We add here a few propositions setting forth some properties of triangles based on similarity and proportion. The first is a remarkable property of right triangles.

XVIII. *If the altitude is drawn on the hypotenuse of a right triangle, then:*

  (1) *The two right triangles formed are similar to the original triangle and to each other;*
  (2) *The altitude is the mean proportional between the segments of the hypotenuse; and*
  (3) *Each leg of the original triangle is the mean proportional between its adjacent segment and the hypotenuse.*

In Fig. 125 let ABC be the right triangle with the right angle at C, and CD the altitude on the hypotenuse AB. We then have to prove

  (1) triangles BCA, CDA and BDC are similar;
  (2) $AD:CD = CD:DB$; (3) $AB:AC = AC:AD$.

For convenient reference mark the two angles at C as shown. Then in right triangles BCA, CDA the acute $\angle A$ is the same. Hence $\triangle CDA$ is similar to $\triangle BCA$ (by XIV above), and also, since in right triangles BCA, BDC acute $\angle B$ is the same, $\triangle BDC$ is similar to $\triangle BCA$. Thus: (1), the smaller triangles are similar to the original and so to each other. —Q.E.D.

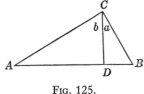

Fig. 125.

Then in the similar triangles BDC, CDA, by definition of similar triangles: (2) $AD:CD = CD:DB$.
                                                                                    —Q.E.D.

Similarly in the pairs of similar triangles BCA, CDA and BCA, BDC:
(3) $AB:AC = AC:AD$ and $AB:BC = BC:BD$.   —Q.E.D.

XIX. *The sum of the squares of the legs of a right triangle equals the square of the hypotenuse.*

To prove this, we have from the results (3) of the preceding proposition, according to the principle II, 61 (referring to Fig. 125),

$$\overline{AC}^2 = \overline{AB} \times \overline{AD}$$
$$\overline{BC}^2 = \overline{AB} \times \overline{BD}$$

Adding, $\overline{AC}^2 + \overline{BC}^2 = \overline{AB} \times \overline{AD} + \overline{AB} \times \overline{BD}$
$$= \overline{AB} \times (\overline{AD} + \overline{BD})$$
$$= \overline{AB} \times \overline{AB}, \text{ since } (\overline{AD} + \overline{BD}) = \overline{AB}.$$

Therefore $\overline{AC}^2 + \overline{BC}^2 = \overline{AB}^2$. —Q.E.D.

This is the celebrated *Pythagorean Theorem*, which is mentioned in article 3 and stated without proof as XXII, 41. It is thought by some historians that the proof given here is the one first given by Pythagoras himself.

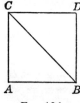

Fig. 126.

This theorem has an interesting application to the square. Thus in Fig. 126 consider the square *ABCD* and its diagonal *BC*. Then since each angle of the square is a right angle and all the sides are equal, the two triangles *BAC*, *BDC* are congruent right triangles, and hence in either triangle, say $\triangle BAC$, $\overline{BC}^2 = \overline{AB}^2 + \overline{AC}^2$. But $AC = AB$; hence $\overline{BC}^2 = \overline{AB}^2 + \overline{AB}^2 = 2\overline{AB}^2$. Dividing the first and last members of this equation by $\overline{AB}^2$ we get $(\overline{BC}^2/\overline{AB}^2) = 2$ or $(\overline{BC}/\overline{AB})^2 = 2$. Hence, taking the square root, $\dfrac{\overline{BC}}{\overline{AB}} = \sqrt{2}$. But the square root of 2 cannot be expressed exactly as a fraction or as a decimal. That is, the ratio of the diagonal and the side cannot be expressed exactly as a fraction or as a decimal. In other words (see article 55),

XX. *The diagonal and the side of a square are incommensurable.*

We shall later find that this is also true of the diameter and circumference of a circle.

65. Circle Proportions. We give here a few of the important proportions among certain lines related to the circle.

XXI. *The perpendicular to the diameter from any point on the circle is the mean proportional between the segments into which it divides the diameter.*

In Fig. 127 let *ACB* be the circle with *AB* as diameter and with *CD* $\perp$ *AB* from any point *C* on the circle. The theorem states that *CD* is **a mean** proportional between the segments *AD* and *BD*.

To prove this, draw AC and BC. Then $\triangle ABC$ is a right triangle, as it is inscribed in a semi-circle (XVIII, 56), and CD is the altitude on the hypotenuse. According to XVIII, 64, therefore, $AD:CD = CD:BD$, and CD is the mean proportional, as was to be proved.

According to VIII, 61, the length of CD is $\overline{CD} = \sqrt{\overline{AD} \times \overline{BD}}$.

On account of this relation a mean proportional is often called a "geometric mean."

XXII. *If two chords intersect in a circle, the product of the segments of one chord equals the product of the segments of the other.*

In Fig. 128 let AB, CD be the chords, intersecting at E. We then have to prove that $\overline{AE} \times \overline{BE} = \overline{CE} \times \overline{DE}$.

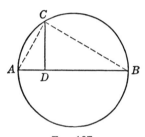
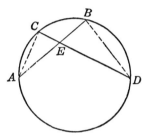

FIG. 127.   FIG. 128.

Draw AC and BD. Then $\angle A = \angle D$, since both are inscribed in $\stackrel{\frown}{BC}$, and similarly $\angle C = \angle B$ (by XX, 56). Also vertical $\angle AEC = \angle BED$. Hence the three angles of $\triangle AEC$ equal those of $\triangle BED$, and the triangles are similar (XII, 64). Their corresponding sides are therefore proportional, by definition of similar figures; that is, $AE:DE = CE:BE$. Therefore, by I, 61, $\overline{AE} \times \overline{BE} = \overline{CE} \times \overline{DE}$, as was to be proved.

XXIII. *If from a point outside a circle a secant and a tangent are drawn, the tangent is the mean proportional between the whole secant and its external segment.*

In Fig. 129 let AB be the secant and AC the tangent from the point A outside the circle. Then AE is the external segment of the secant and the proposition states that $AB:AC = AC:AE$.

Draw BC and CE. Then $\angle B = \angle ACE$, for each is measured by $\frac{1}{2}\stackrel{\frown}{CE}$ (XVII, XXI, 56). Also $\angle A$ is common to both triangles ABC,

ACE. Hence triangles *ABC*, *ACE* are similar (XIII, 64). Therefore $AB:AC = AC:AE$, as was to be proved.

XXIV. *If a triangle is inscribed in a circle, the product of any two sides equals the product of the altitude on the third side and the diameter.*

In Fig. 130 let *ABC* be the inscribed triangle, *CD* the altitude on side *AB*, and *CE* a diameter of the circle. The proposition then states that $\overline{CA} \times \overline{CB} = \overline{CD} \times \overline{CE}$.

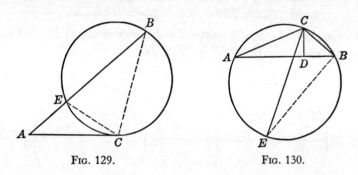

Fig. 129.   Fig. 130.

Draw *BE*. Then since *CE* is a diameter △*CBE* is a right triangle; and so is △*CDA*. Also acute ∠*CAB* = ∠*CEB*, since both are measured by $\frac{1}{2}\overset{\frown}{BC}$. Hence the right triangles *CBE*, *CDA* are similar (XVI, 64). Therefore, by definition of similar triangles, $CA:CE = CD:CB$, and hence $\overline{CA} \times \overline{CB} = \overline{CD} \times \overline{CE}$, as was to be proved.

There are many other interesting proportional relations among the lines associated with circles but those we have given are sufficient to illustrate the methods of proof to be used.

66. Construction Problems.

XXV. *To divide a given straight line into parts proportional to any number of given line segments.*

In Fig. 131 let *l*, *m*, *n* be the given segments and *AB* the line to be divided proportionally to *l*, *m*, *n*.

Through *A* draw *AC* of any convenient length and at any convenient angle with *AB*, and on *AC* lay off $AD = l$, $DE = m$, $EF = n$. Join *BF* and through *D*, *E* draw $DG \parallel EH \parallel BF$.

G and *H* are then the required division points on *AB*, making *AG*, *GH*, *HB* proportional to *l*, *m*, *n*.

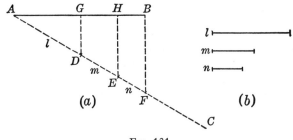

FIG. 131.

For, ABF is a triangle with DG and HE parallel to its base BF and hence, according to XI, 62, $AG:AD = GH:DE = HB:EF$, or putting in l, m, n for their equals, $AG:l = GH:m = HB:n$, as was required.

XXVI. *To find the fourth proportional to three given line segments.*

In Fig. 132(a) let l, m, n be the three line segments, and in (b) draw AB and AC at any convenient angle. On one line in (b), say AB, lay off $AD = l$, $DE = m$, and on AC lay off $AF = n$. Then join DF and through E draw $EG \parallel DF$, meeting AC at G. Then FG is the required fourth proportional.

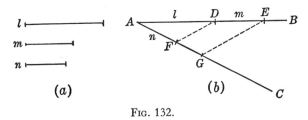

FIG. 132.

For, according to the construction and IX, 62,

$$AD:DE = AF:FG,$$

or, substituting equals, $l:m = n:\overline{FG}$ and FG is the fourth proportional to l, m and n.

XXVII. *To find the third proportional to two given line segments.*

The solution is the same as that of XXVI preceding, except that l and m are given and $AF = DE = m$ also, instead of n. Then $l:m = m:\overline{FG}$ and FG is the third proportional to l and m.

XXVIII. *To find the mean proportional to two given line segments.*

In Fig. 133(a) let m and n be the given segments, and in (b) draw AB equal in length to $m + n$, with $AD = m$, $BD = n$. On AB as a diameter describe a semi-circle and at D erect $DC \perp AB$ meeting the semi-circle at C. Then CD is the required mean proportional.

For, according to XXI, 65, $DA:DC = DC:DB$, or $m:DC = DC:n$, and hence DC is the mean proportional to m and n.

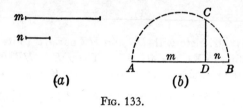

Fig. 133.

67. Illustrations and Applications. The number of illustrations and applications of proportion and similar figures in the arts, sciences, engineering, industry, daily life and observation, etc., is unlimited, as indeed is the case with every division of geometry. On account of space limitations we shall give here only a few, sufficient to bring out the interest and full meaning of a few of the principles developed in this chapter.

(i) *Shadows and Enlarged Pictures.* These are familiar examples of similar figures. If a small source of light shines perpendicularly on a flat surface, as a wall or screen, and an object is placed between the surface and the light, the shadow on the surface will be of exactly the same shape as the outline of the object as presented to the light, but larger than the object.

A ray of light from the source passes every point of the outline of the object and reaches to the surface in a straight line. Inside this set of lines no rays pass the object and the surface is not illuminated inside the figure of that outline. The shadow is said to be *cast* by these outline rays, or by the light.

If the object is moved nearer the surface, the shadow is smaller, and if moved nearer the light, the shadow is larger, each part of the outline being increased in the same ratio, while the *shape* (angles) remains the same. The shadow is thus *similar* to the outline of the object.

Enlarged pictures and telescope and microscope images are produced in a manner analogous to that in which shadows are produced and are also *similar* to the object or original.

"Similar" as here used has the geometrical meaning. As regards visual appearance it may be said in ordinary language that the shadow, image or picture is "exactly like" the original.

(ii) *The Pantograph*. This is an instrument used to reproduce drawings and sketches on a larger or smaller scale by tracing the lines of the figure with a pointer which is attached to or controls a lever or beam which in turn carries a pencil or pen for making a geometrically *similar* figure on another sheet.

One form of pantograph is shown in Fig. 134. It consists of four straight bars, parallel in pairs and jointed at B, C, D and E. At D is a tracing point and at F a pencil, while at A is a fixed pivot. By shifting the pivot joints B and E the quadrilateral $BCDE$ can be given any desired shape and size. When D traces out any figure, such as

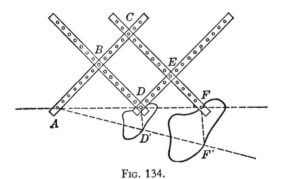

Fig. 134.

that shown, F traces out a similar but larger figure, of any desired size. By interchanging the tracer D and pencil F any figure traced by F is reproduced on a smaller scale by D.

The principle of the instrument is simple. The pins at B, E are set so that $BCDE$ is a parallelogram and the points B, E divide the arms AC, CF in any desired proportion $\frac{AB}{AC} = \frac{CE}{CF}$. Then ADF remains always a straight line no matter how the line turns about A, and as D moves sidewise or endwise of the line AF, F moves in the same direction but farther.

This is due to the fact that in any position of ADF, $AD:AF = AB:AC$

and also, in any position D', F', $DD':FF' = AD:AF$, the ratio $AD:AC = CE:CF$ being that set by the adjustment.

It is easily seen that these proportions hold. For, since $BCDE$ is a parallelogram, $BD = CE$, and the adjustment proportion becomes $AB:AC = BD:CF$, and therefore the triangles ABD, ACF are similar (sides proportional). Therefore, in the same triangles, $AD:AF = AB:AC$; and also in another position $AD':AF' = AB:AC$. Hence the triangles ADD', AFF' are similar, and therefore $DD':FF' = AD:AF$, or $DD':FF' = AB:AC$.

The pantograph was invented in 1603 by Christopher Scheiner and is very widely used by draftsmen and mapmakers. Fig. 134 shows only one of many forms and combinations.

(iii) *Measuring Inaccessible Distances.* This is an old and important problem to surveyors, builders and mapmakers, as well as one of interest and frequent importance in everyday life. Several methods have already been given (article 46). Those given here make use of the properties of similar figures.

In Fig. 135(*a*) let AB represent a pond, swamp, hill or other irregular shape such that the distance AB cannot be measured directly. Sight

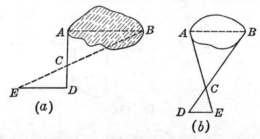

Fig. 135.

the line AB and set two stakes at convenient points D, E so that $AB \perp AD \perp DE$. Sight along EB and note where the line of sight crosses AD, at C. Then triangles BAC, CDE are right triangles, and acute $\angle ACB = \angle DCE$ (as vertical angles). Therefore right $\triangle BAC$ is similar to $\triangle CDE$ (XIV, 64), and hence $AB:AC = DE:DC$. Therefore the product of the extremes $\overline{AB} \times \overline{DC} = \overline{AC} \times \overline{DE}$ the product of the means, and hence $AB = \dfrac{\overline{AC} \times \overline{DE}}{\overline{DC}}$. As AC, DE and DC are easily measured, AB is calculated at once.

The same problem is also easily solved by the method indicated in Fig. 135(b). Here no angles have to be laid off but two additional distances have to be measured. This is easy, however, with a tape line.

From C sight on A and on B and measure AC, BC. Then extend these lines and mark D and E so that $CA:CE = CB:CD$, and measure DE. Then $DE:AB = CD:CB$ and hence $AB = \overline{CB} \times \overline{DE} \div \overline{CD}$.

This result depends on the similarity of triangles ABC and CDE. For, as laid out $CA:CE = CB:CD$, and vertical $\angle ACB = \angle DCE$ (XV, 64). In these similar triangles then $DE:AB = CD:CB$, as stated.

The distance from A to the inaccessible point P in Fig. 136(a) is easily measured by running a line from A to C, in line with AP, and laying off two lines perpendicular to PC; one from A to a convenient point B, and one from C to a point D in line with PB, and then running $BE \perp CD$.

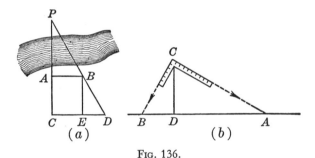

Fig. 136.

Then $AB \parallel CD$ (both perpendicular to PV), and $BE \parallel PC$ (both perpendicular to CD). The right triangles PAB, BED then have the same angles and hence are similar. Therefore $AP:AB = BE:ED$ and hence $AP = \overline{AB} \times \overline{BE} \div \overline{ED}$.

After laying off the perpendiculars AB, CD, BE (by methods already described in article 43), it is therefore only necessary to measure AB, BE and ED and the formula gives AP at once.

Suppose the distance DA in Fig. 136(b) is required and that A is inaccessible from D. DA is easily found by a simple use of the carpenter's square or a similar square cut from cardboard, or made of two sticks or a bent wire making a right angle.

At D drive a vertical stake ($\perp AB$), measure the height CD, and hang the square on the stake as shown. Sight along one edge of the square at A, and with the square held in the same position sight along the other edge at the level ground, marking the point B; then measure BD.

Triangle ABC is a right triangle and CD is the altitude on the hypotenuse AB. Therefore CD is the mean proportional between AD and BD (XVIII(2), 64). That is, $AD:CD = CD:BD$, and hence $AD = \overline{CD}^2 \div \overline{BD}$.

(iv) *Measuring Inaccessible Heights.* The measurement of inaccessible heights is a simple problem when the properties of similar right triangles are utilized. The ancient Greek business man and philosopher Thales (article 3) is said to have astonished the Egyptians by determining the height of one of their largest pyramids simply by making one measurement on level ground.

The method of Thales consisted simply in noting the time when the length of his shadow equaled his height and marking at that moment the tip of the shadow of the pyramid. The length of the shadow was then measured, and was of course the same as the height of the pyramid.

This method is explained by reference to Fig. 137(*a*). Here PCD represents the pyramid and AP its height. AB is the shadow and at

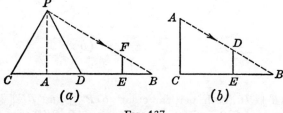

Fig. 137.

the same time EB is the shadow of the man EF. In the similar right triangles BEF and BAP, when $EB = EF$ then also $BA = AP$. AD is inside the base of the pyramid, but is parallel to the outer edge of the base and so can be measured, and most of the pyramids of Egypt are so placed that this edge is very nearly parallel to the direction of the shadow.

A modification of this method is shown in 137(*b*) and applies at any time, and not only when the sun is at a particular elevation. Thus let CA be a height which is to be determined and CB its shadow at any time. If at the same time BE is the shadow cast by a vertical stick or pole DE of known length, then the right triangles BED and BCA are similar, and hence $AC:BC = DE:BE$. From this we have at once $AC = \overline{BC} \times \overline{DE} \div \overline{BE}$. By measuring the shadows BC and BE therefore, when the height DE is known, the height CA is found at once.

Another simple method for measuring heights is illustrated in Fig. 138(a). The rectangle $EABC$ is cut from cardboard, wood or metal and graduated as a ruler on two adjacent edges, and AP is a plumb bob fastened at A. T is the top of a tall object, as a tree or building; FG is the level ground, and FE represents the height of the person making the measurement.

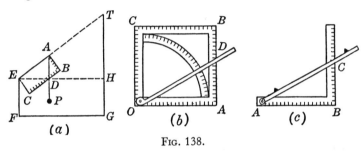

Fig. 138.

The rectangle is held with one corner E at the eye of the observer, who stands at F and sights along the edge EA at T, and notes the point D where the free plumb line crosses the edge BC. The distance FG is then measured, and from this and the known lengths AB and BD the height GT is found, as follows.

As two sides of a rectangle $BA \perp AE$, and also the vertical $AD \perp EH$ the horizontal; that is, the sides of acute $\angle BAD$ are perpendicular to those of $\angle HET$. Hence the two angles are equal, and therefore the right triangles BAD and HET are similar.

From these similar triangles we have the proportion $HT:HE = BD:AB$, and therefore $HT = \overline{HE} \times \overline{BD} \div \overline{AB}$. Since AB and BD are known this formula gives HT. When to this is added the height of the observer $EF = HG$, the result is the height GT. The complete formula is therefore $GT = \overline{FG} \times \overline{BD} \div \overline{AB} + \overline{EF}$.

Several other simple instruments which utilize the properties of similar right triangles are used to measure heights. Two of these are shown in Fig. 138(b) and (c).

The sketch in (*b*) represents the *quadrant*, an instrument which was standard with surveyors and astronomers before the invention of the telescope and *transit*, and which is still useful for many quick and simple measurements.

The frame $OABC$ is a square with edges graduated as rulers, AC is a quadrant of a circle graduated in degrees like a protractor, and OD is a straight arm which rotates about O. If OA is held horizontal and AB vertical and a sight is taken along OD, then $\angle AOD$ is read on the quadrant and OA, AD are read off as the legs of a right triangle. This triangle is similar to the one formed by the distances or heights involved in the measurement, and hence these are found from the proportions worked out in the preceding examples.

The sketch in (*c*) represents an instrument called the *hypsometer*, which is sold regularly by dealers for measuring heights and distances. It is of the form shown and is made of brass with each arm AB, BC equally divided into, say, 100 divisions, which are read on the scales, and at the ends of the hinged radius arm AC are sights for aiming. If the observer stands 100 feet from a tree, for example, and sights along AC at the top of the tree while the arm AB is held horizontal, then if BC reads, say, 75 the tree is 75 feet high.

(v) *Finding Diameter of a Circle.* Suppose it is required to find the diameter of a circle already drawn; how is this to be done? It may seem at first thought that a ruler can be laid across the center and the diameter measured. But the center may not be known. Also the circle may be the circumference of a round rod or post, etc., as well as a circle on a drawing. In such cases the center is not accessible and the cross section of the post may not be accessible for measurement.

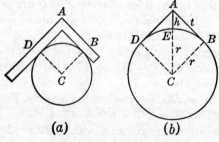

Fig. 139.

ART. 67 PROPORTION AND SIMILAR FIGURES 153

In such cases several methods are available, depending on the proportions of article 66 and other properties of the circle already studied.

We give here several methods, two available when the center is unknown or the interior is not accessible (Fig. 139), and others applicable when the whole interior or part of it is accessible but the center unknown (Fig. 140).

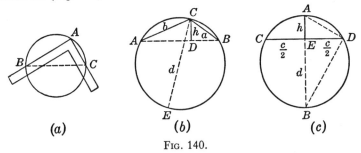

FIG. 140.

The method of Fig. 139(a) is suitable for a small rigid material circle, such as a pole or rod. Lay a carpenter's square or other right angle against the circle as shown. Then, since the angles at B, D are right angles (X, 54) and $\angle A$ is a right angle, $ABCD$ is a square, where C is the center and $CD = CB$ the radius. Therefore $AB = AD = CD = CB$ and the radius is read off or measured on the square at AB. The diameter is then $2 \times \overline{AB}$.

If the circle is too large for the method of (a) or is drawn on a sheet, the method of (b) will apply. From any convenient outside point A draw tangents to B, D, or lay straight edges AB, AD against the circle, and bisect $\angle BAD$. Measure AB and on the bisector measure $AE = h$, and let the radius $BC = r$.

Then $AB = AD = t$, say (XI, 54), and $\angle B$ is a right angle. Therefore $\triangle ABC$ is a right triangle and hence $\overline{AC}^2 = \overline{AB}^2 + \overline{BC}^2$. That is, $(r + h)^2 = r^2 + t^2$. From this, by expanding and simplifying,

$$r^2 + 2rh + h^2 = r^2 + t^2$$

$$2rh + h^2 = t^2$$

$$2rh = t^2 - h^2$$

$$2r = \frac{t^2 - h^2}{h}$$

But $2r = d$, the diameter, and $t^2 - h^2 = (t+h)(t-h)$. Therefore,
$$d = \frac{(t+h)(t-h)}{h}.$$

This says, in words: add t and h, and also subtract; multiply the sum by the remainder, and divide the product by h. The quotient is d.

The method of Fig. 140(a) serves for a small circle and is based upon the fact that a right angle is inscribed in a semi-circle. Lay a draftsman's triangle, carpenter's square or piece of square-cornered cardboard over the circle so that the vertex of the right angle is on the circle, and mark the points B, C where the edges cross the circle. Then BC is the diameter.

In (b) let $AEBC$ be the circle and inscribe any triangle ABC. Measure the sides $AC = b$, $BC = a$, and the altitude (perpendicular) $CD = h$. Then according to XXV, 66, $ab = hd$, and hence $d = ab/h = CE$ is the diameter.

In (c) let $ABCD$ be the circle; draw and measure any chord $CD = c$, and draw and measure its perpendicular bisector $AE = h$ (called the *rise* of arc $\overparen{CAD}$).

Then $AB = d$ is the diameter, $\triangle ADB$ is a right triangle, and DE is its altitude. According to XVIII(3), 64, therefore, $h:AD = AD:d$ and hence $\overline{AD}^2 = dh$. Also $\triangle AED$ is a right triangle and hence

$$\overline{AD}^2 = h^2 + \left(\frac{c}{2}\right)^2 = h^2 + \tfrac{1}{4}c^2.$$

Putting these two expressions for $\overline{AD}^2$ equal to one another, $dh = h^2 + \tfrac{1}{4}c^2$, and therefore

$$d = h + \frac{c^2}{4h}.$$

This says that the diameter is equal to the rise of the arc plus the quotient of the chord squared by four times the rise.

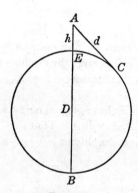

FIG. 141.

(vi) *Size of the Earth and Distances on the Earth.* One method of finding the size of the earth, by means of its circumference, is given in article 34. We give here a method which uses the diameter.

Let Fig. 141 represent the earth, $\overline{BE} = D$ its diameter, $\overline{EA} = h$ a height above the earth, and $\overline{AC} = d$ the distance from A to some distant point C on the horizon.

Then AC is a tangent, the line AB through the center is a secant, and AE is the external segment of the secant. According to XXIII, 65, therefore, $AB:AC = AC:AE$. Using the symbols given in the figure, and noting that $\overline{AB} = D + h$, the proportion written as an equation becomes $\dfrac{D+h}{d} = \dfrac{d}{h}$. Hence $D + h = \dfrac{d^2}{h}$ and $D = \dfrac{d^2}{h} - h$.

Thus if the height h is measured, as in some of the examples given in the first part of this article, and C is an object on the horizon (say at sea) at a known distance d from A the top of the height, the difference between the height and the quotient of the distance squared by the height, gives the diameter of the earth. By such methods the diameter is found to be nearly eight thousand miles, more nearly 7920.

Suppose on the other hand, that the size of the earth is known. Then from any particular height above the earth, what is the greatest distance at which an object on the earth can be seen? This question is answered by solving the formula just found for d, considering D and h as known. Thus the original proportion is $\dfrac{(D+h)}{d} = \dfrac{d}{h}$. Hence by II, 61, $d = \sqrt{(D+h)h}$.

Since D is nearly eight thousand miles and any height h above the earth is at most only a few miles, then h is so small in comparison with D that for all ordinary calculations we can use D instead of $D + h$. The distance formula can then be written in the very simple form

$$d = \sqrt{Dh}.$$

Thus from the top of the Empire State Building in New York City ($h = 1250$ feet $= .237$ mile) one can see over the surrounding country and ocean to a distance of $d = \sqrt{7920 \times .237} = \sqrt{1877} = 43.3$ miles. Of course this is affected somewhat by the atmosphere, and at this distance objects will not be distinct without a telescope, but the country is exposed to view to this distance on every side.

An aviator at a height of five miles above the earth can see as far as $d = \sqrt{7920 \times 5} = \sqrt{39,600} = 199$ miles, very nearly two hundred miles.

68. Exercises and Problems.

1. What are the whole, mixed or decimal number values of the following ratios: $\tfrac{8}{4}$, 9:6, 9/4, $\tfrac{15}{20}$?

2. What is the simplest form of expression of the ratio of 50 feet to 125 feet?

3. If the three segments cut out of an oblique transversal by four parallels are 2, 3, 4 inches and the perpendicular distance between the first pair of parallels is one inch, what are the distances between the other parallels?

4. How could the page of a ruled notebook be divided accurately into three equal columns without measuring?

5. Given an angle ABC and a point P lying anywhere between the sides. Construct a line through P such that the segment EF of it between the sides of the angle is bisected by P. Prove the construction.

6. Find by geometrical construction and check by calculation the fourth proportional to 2, 3 and 8; and the third proportional to 1 and 4.

7. Draw any triangle and enlarge its sides in the ratio 3:5 by geometrical construction alone.

8. How high is a chimney whose shadow is 144 feet long when a five-foot gate post casts a twelve-foot shadow?

9. Prove as a theorem that the diagonals of a trapezoid divide each other in the same proportion.

10. Let P, Q be any two points on the sides of an angle ABC, and draw perpendiculars from P and Q to each side opposite. Prove that the triangles so formed are similar.

11. What is the distinction between *similar* and *congruent*?

12. The altitude on the hypotenuse of a certain right triangle divides the hypotenuse into segments of 4 and 16 inches. How long is the altitude? How long is it when the segments are each 10 inches?

13. If one of the legs of a certain right triangle is three times the other, in what ratio does the altitude on the hypotenuse divide the hypotenuse?

14. A chord of a circle is divided into two segments of 3 and 5 inches by another chord, one of whose segments is 4 inches. How long is the second chord?

15. A secant is drawn from a point outside a circle in such a way that its whole length is 9 inches and the segment cut off inside the circle is 4 inches. What is the length of the tangent to the circle from the same point?

16. A ruler laid across the edge of a piece of a broken wheel (as a chord) shows 14 inches from edge to edge, and the perpendicular from the edge to the middle point of the chord is 3 inches. What was the diameter of the wheel?

17. The Eiffel Tower in Paris is about 0.186 mile high. How far can one see from the top of the tower?

18. An airplane is disabled by a shell hit when it is 6500 feet up above level country, and at once begins to glide downward in a straight line toward the ground. It lands at a point two miles from the point immediately below where it was hit. How far does the plane glide?

19. Just as an airplane passes over a village an observer stationed $3\frac{1}{2}$ miles from the village in level country gets a sight on the plane with a range-finder and determines that the plane is $4\frac{1}{2}$ miles from him in a straight line. How high up is the plane above the village?

20. As an airplane just clears a mountain peak known to be 12,500 feet high the pilot sights a distant target on a range-finder and finds that it is about six miles from him in a straight line. If the target is on the level with the foot of the mountain, how far must the plane fly to be just over the target at the same height?

21. How far over the surrounding country can an airplane pilot see with glasses when he is flying three miles above sea level?

Part III

MEASUREMENT
OF GEOMETRICAL FIGURES

Chapter 7

DIMENSIONS AND AREAS OF PLANE FIGURES

69. Introduction. In the preceding chapters, plane figures have been considered mainly from the viewpoint of the number of their sides and angles and relations between their shapes, and the methods of constructing them. In the comparison of angles, however, it was necessary to consider their *measure* also, and in the study of proportion considerable attention was paid to the measure of lines, that is, the quantity called *length*.

In the chapter on proportion and similarity, the lengths of lines are considered from the viewpoint of comparison, the comparison of the lengths of the lines of one figure with the lengths of those of another figure (their proportion) or of other lines in the same figure, e.g., in the case of the right triangle ratios.

We shall now consider the lengths of the lines forming a plane figure as constituting in themselves certain important properties of the figure *itself*, without reference to any other.

In this sense the lengths of the straight lines forming a plane figure, or of certain other related lines of the figure, are referred to as its *dimensions*, which we now consider in detail.

70. Dimensions of Plane Figures. In article 15 the three lengths, or lengths in the three directions of space, which are necessary to describe a solid are said to be the three *dimensions* of the solid. Since a surface has no thickness it has only two dimensions. The surface in which we have been most interested is the plane, and in the sense of article 15 a plane and also any plane figure has two and only two dimensions.

For our present purpose, however, we must be more explicit. A dimension of a plane figure will therefore be taken as the *measure of the length* of a straight line segment which is a side or other line associated with the figure. In particular,

The *dimensions of a plane figure* are the lengths of those straight lines of the figure which must be given in order to describe its shape and size.

Thus the dimensions of a rectangle are its *length* and *width* (base and altitude) or either one of these and its *diagonal*. The dimensions of a triangle are its *sides* or its sides and *altitudes;* those of a circle are its *radius, diameter* or *circumference;* etc.

In the following discussions of the measurement of various plane figures, the dimensions of each will be specified.

71. Area and Its Measure. In all our study of plane figures so far, a plane figure has been thought of only as made up of straight or curved lines; in fact the lines not only form the figure, they *are* the figure, and a plane figure *is* its lines.

There is another thing to be considered in connection with plane figures, however: What of that *portion of the plane enclosed* by the figure? Certainly some plane figures enclose greater portions of a plane than others. How can we *compare* the plane portion enclosed by one figure with that enclosed by another? How can we describe the plane portion enclosed by any *one* figure, and is there a way to *measure* it? These questions all admit of definite answers. We begin with the following

DEFINITION: *The* AREA *of a plane figure is the numerical measure of the portion of its plane enclosed by it.*

In order to express the numerical measure of an area, or of any quantity, we must first have a *unit of measure* for that quantity (article 55). Thus the unit of measure of length is the inch, foot, yard or mile in the *English* system; in the *Metric* (decimal) system it is the centimeter, meter or kilometer.

The unit of length having been selected, the unit of area is then defined as follows: *The unit of area is a square whose side is a unit of length.* This unit is named by placing the word "square" before the unit of length; thus the *square* inch, *square* centimeter, *square* foot, etc.

Having the unit of area we then say (as in article 55) that *the area of a plane figure is the number of units of area contained in the portion of plane enclosed by the figure.*

Thus, if the size of a rectangle is such that fifteen unit squares, say the square inch, can be placed inside it, edge to edge and without overlapping or leaving any of the enclosed plane portion uncovered, the area of the rectangle is 15 square inches.

If two plane figures have the same area they are said to be *equal*.

Equal figures must not be confused with congruent figures. Thus plane figures are *congruent* when one can be placed on the other so that they coincide throughout, line by line, angle by angle and point by point. *Two figures may be equal without being congruent.* Thus a triangle and a quadrilateral may have the same area, but they can never be fitted together and be made to coincide. Figures which are congruent, however, obviously are also equal.

We next develop some propositions and rules relating to the areas of various plane figures and the relations of these areas to the dimensions of the figures.

72. Areas of Rectangles and Parallelograms. Before developing the rules for calculated areas we must first see how areas are *compared*. We begin with the proposition

I. *The ratio of the areas of two rectangles having equal altitudes equals the ratios of their bases.*

In Fig. 142 let $ABCD$, $AEFD$ be the two rectangles to be compared, having bases AB, AE and equal altitudes AD. In referring to rectangles we shall use the abbreviation $\square AC$ to mean "rectangle $ABCD$," naming the letters of diagonally opposite vertices and using the symbol $\square$ to mean "rectangle."

Our theorem then states that

$$\frac{\text{area } \square AC}{\text{area } \square AF} = \frac{AB}{AE},$$

and this can be proved.

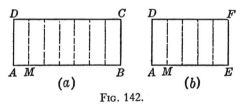

Fig. 142.

We can also prove

II. *The ratio of the areas of any two rectangles having equal bases equals the ratio of their altitudes.*

By combining these two results we can prove the following very important third proposition:

III. *The ratio of the areas of two rectangles of any bases and altitudes equals the ratio of their products of base and altitude.*

Let $ABCD$, $EFGH$ (Fig. 143) be any two rectangles, and let the measure of their altitudes be a, a'; their bases b, b'; and their areas A, A' (in corresponding length and area units). The theorem then states that

$$\frac{A}{A'} = \frac{a \times b}{a' \times b'},$$

and this is to be proved.

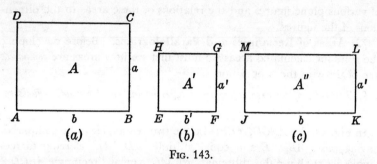

Fig. 143.

Construct a rectangle $JKLM$ having a base equal to $AB = b$ and an altitude equal to $FG = a'$, and an area A'', as in (c).

Then, comparing rectangles AC and JL as in proposition I above, we have $A/A'' = a/a'$, and comparing rectangles JL, EG as in II we have $A''/A' = b/b'$.

Multiplying the first of these equal ratios together, and also the second, we get

$$\frac{A \times A''}{A'' \times A'} = \frac{a \times b}{a' \times b'},$$

according to the equality multiplication axiom. Cancelling A'' in the fraction on the left of this equation,

$$\frac{A}{A'} = \frac{a \times b}{a' \times b'},$$

as was to be proved.

This result leads directly to the following extremely important theorem or rule:

IV. *The area of a rectangle equals the product of its base and altitude.*

In Fig. 144(a) let $ABCD$ be the given rectangle, $AB = b$ its base, $BC = a$ its altitude, and A its area. We then have to prove that $A = a \times b$.

ART. 72 DIMENSIONS AND AREAS OF PLANE FIGURES   165

In (b) let U be a square of side 1 unit of length. Its area is then the unit of area, $U = 1$ square unit. That is, the unit square is a rectangle of base $b' = 1$ and altitude $a' = 1$.

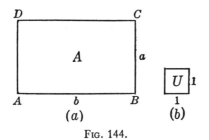

Fig. 144.

According to the proposition III above, therefore,

$$\frac{A}{U} = \frac{a \times b}{1 \times 1}.$$

But $U = 1$ and $1 \times 1 = 1$. Therefore,

$$A = a \times b$$

as was to be proved.

This principle or rule is illustrated and explained in arithmetic, but not proved. We have here obtained a complete proof of it, based logically on principles already established, according to the formal methods of geometry.

The proposition IV is the fundamental theorem or rule of the whole theory of areas. It can be stated in several forms. Thus *for any rectangle, the area is*

(base) × (altitude), $A = b \times a$;
(length) × (width), $A = L \times W$;
(length) × (breadth), $A = L \times B$.

When writing the product of numbers, as 5×16, the multiplication sign is necessary but in indicating the product of algebraic symbols the sign is ordinarily not used. Thus $a \times b = ab$. Therefore for any rectangle

$$A = ab, \quad A = LW, \quad A = BL$$

are equivalent forms of the *area formula*.

Now the altitude and base, or the length and width (or breadth) of a rectangle are its *dimensions*, since they determine its shape, size and area. Therefore *the area of a rectangle is the product of its dimensions*.

The dimensions of a rectangle are usually expressed as a product. Thus a rectangle 5 inches long and 3 inches wide is said to be "5 inches by 3 inches" and this is written 5 in. × 3 in., or when there is no chance of confusion of the sign with seconds of angle, as 5″ × 3″, and this is generally read as "five by three." The area of this rectangle is of course 5 in. × 3 in. = 15 *square inches*, or, as it is usually written, 15 sq. in. This may sometimes be written as in.² or □″.

Since a square is simply a rectangle whose base and altitude are equal, the area of a square is, by the rule obtained above, the product of a side by itself, $A = a \times a = a^2$, that is, the *algebraic* square of its side. That is,

V. *The area of a square is the algebraic square of its side.*

It is for this reason that in algebra the product of a number by itself is called its *square*.

From the rule for the area of the rectangle we easily derive that for the parallelogram. Thus in Fig. 145(a) let $ABCD$ be a parallelogram of base $AB = CD = b$ and altitude $a = DE \perp AB$.

Let Fig. 145(b) represent the parallelogram of (a) with the triangular section ADE at the left removed and replaced at BCF at the right.

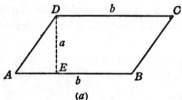 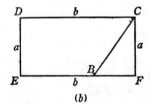

(a)   (b)

Fig. 145.

Then, since the altitude $DE \perp AB$ the base, triangles ADE, BCF are right triangles. Also, as opposite sides of the parallelogram $AD = BC$ (XVIII, 46), and of course the altitude $DE = CF$. Also, since $CF \parallel DE$ and $AD \parallel BC$ as opposite sides of a parallelogram, then acute $\angle ADE = \angle BCF$ (XXI, 32). Therefore $\triangle ADE$ is congruent and so also equal to $\triangle BCF$ (XXI, 41). That is, the same area is added to the parallelogram on the right as was taken off at the left and the figure $CDEF$ has the same area as the parallelogram $ABCD$. But the figure $CDEF$ is a rectangle (sides $\parallel$ and $\perp$) and has the same base and altitude b, a as the parallelogram $ABCD$, and the area of $\square CDEF =ab$, by IV above. Therefore the area of parallelogram $ABCD$ is also equal to ab. That is,

VI. *The area of a parallelogram equals the product of its base and altitude.*

The altitude $a = ED$, Fig. 145(a), which is *perpendicular* to the bases of the parallelogram, must not be confused with the *slant* height AD. Two parallelograms may have the same bases and slant heights but different altitudes.

The base, altitude and slant height completely determine the shape, size and area of a parallelogram and are therefore its *dimensions*.

73. Remarks on Area Measure. Area measure has been defined in terms of the unit of area, and this unit is a square. It is easy to see how the area of a rectangle can be expressed in terms of this square unit, as is explained in arithmetic and illustrated in Fig. 146. Thus if the length is, say, exactly 5 whole units and the width is exactly 3 whole units, then the rectangle is easily divided into 15 unit squares. There are three rows of five each, or five columns of three each. If there is no common measure for both base and altitude, the rule is still seen to be true by the method of taking the unit of measure indefinitely smaller and passing to the limit, as in the proof of proposition I above and as explained in article 55.

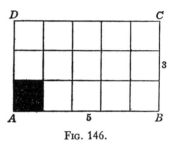

Fig. 146.

It is not so readily seen, however, that a parallelogram with no square corner, or a triangle, can be divided into unit squares, for no *square* unit, no matter how small, can be fitted in between the sides near a vertex. We can easily get around this difficulty, however, as shown above.

74. Areas of Triangles. It is an easy matter to see now how the area of a triangle is to be found from its dimensions. For any triangle whatever can be thought of as half of some parallelogram, as in Fig. 147, where the triangle ABC is half of the parallelogram $ABCE$ (XXIX, 46). But the area of the parallelogram is the product of its base AB by its altitude CD; and the base and altitude of the parallelogram are the same as those of the triangle, $AB = b$ and $CD = h$. Therefore the area of parallelogram $ABCD = b \times h$ and area $\triangle ABC = \frac{1}{2}(b \times h) = \frac{bh}{2}$. That is,

VII. *The area of a triangle equals half the product of its base and altitude.*

As already made clear, the area of a triangle is expressed in the usual square units of area.

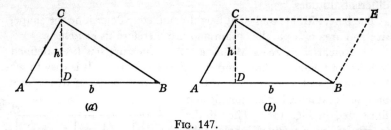

FIG. 147.

If two triangles have equal bases and equal altitudes, then the half product of base and altitude will be the same for both. Therefore

VIII. *Triangles with equal bases and equal altitudes have equal areas.*

Such triangles need not have the same sides, or shape.

If two triangles have equal bases but the altitude of one is three times that of the other, then the area of one is three times that of the other, for the base-altitude product will be three times greater for that triangle; and similarly for equal altitudes and different bases. Therefore,

IX. (1) *The areas of triangles of equal bases are in the same ratio as their altitudes;* and
    (2) *the areas of triangles of equal altitudes are in the same ratio as their bases.*

(Compare propositions I and II above.)

Although it is one side and the altitude on that side of a triangle which determine its area rather than the three sides, the three sides of a triangle are more often given than the altitudes. The question thus arises as to whether and how the area can be found from the sides alone. A formula can be derived which gives the area in terms of the sides, but the derivation of this formula is rather involved and is best

Art. 74 DIMENSIONS AND AREAS OF PLANE FIGURES   169

understood after studying some of the principles of *trigonometry*. It is therefore studied in that subject, and will not be derived here. The formula is easily stated and used, however, as follows. Let the letter s represent half the sum of the three sides, that is, $s = \frac{1}{2}(a + b + c)$. Then subtract the value of each side from this number s, that is, find the values of the three remainders $s - a$, $s - b$, and $s - c$. The area of the triangle is then the square root of the product of these three remainders and s.

Stated concisely, for reference, therefore,

X. *If the sides of any triangle are a, b, c and half the sum of the sides is indicated by s, the area of the triangle is equal to* $\sqrt{s(s - a)(s - b)(s - c)}$.

As an example consider the triangle whose sides are $a, b, c = 3, 4, 5$ inches respectively. Then, $s = \frac{1}{2}(3 + 4 + 5) = 6$; $s - a = 3$, $s - b = 2$, $s - c = 1$, and the area is $\sqrt{6 \times 3 \times 2 \times 1} = \sqrt{36} = 6$ sq. in.

This formula for the area of any triangle is very easily adapted to give very simple formulas for isosceles and equilateral triangles.

For example, it can be shown that the area of an equilateral triangle is

$$A = \frac{a^2}{4}\sqrt{3}.$$

Thus an equilateral triangle 6 feet on a side has an area

$$A = \frac{6^2}{4}\sqrt{3} = 9\sqrt{3} = 15.59 \text{ sq. ft.}$$

For convenient reference these results are stated here as

XI. (1) *If the legs of an isosceles triangle are a and the base c the area is* $A = \frac{1}{4}c\sqrt{4a^2 - c^2}$.
(2) *If the side of an equilateral triangle is a the area is* $A = \frac{1}{4}a^2\sqrt{3} = 0.433a^2$.

In the case of a right triangle, two of the sides are perpendicular, and these two are called the legs. This means that if one of the legs be taken as base the other is the altitude of the triangle. Therefore

XII. *The area of a right triangle is half the product of its legs.*

If a diagonal of a rectangle of base b and altitude a is drawn, the rectangle, which is a parallelogram, is divided into two equal right triangles whose legs are a and b. And since the area of the rectangle is ab the area of the right triangle is $\frac{1}{2}ab$, half the product of its legs, as before.

Consider the triangle having sides $a = 3$, $b = 4$, $c = 5$ inches. This is a right triangle because it satisfies the *Pythagorean Theorem* $a^2 + b^2 = c^2$, $3^2 + 4^2 = 5^2$. Therefore $a = 3$ and $b = 4$ are the legs and $c = 5$ the hypotenuse. The area is therefore $A = \frac{1}{2}ab = \frac{1}{2}(3 \times 4) = 6$ sq. in. This is the same triangle that was used as an example to illustrate the formula of X above. That formula is therefore checked in this case.

We give next the demonstration of two very interesting propositions concerning the ratios of triangle areas, which are very simple. The first is

XIII. *If two triangles have one angle equal, their areas are in the same ratio as the products of the sides containing the equal angles.*

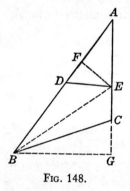

Fig. 148.

Since an angle of one triangle equals an angle of the other, one of the triangles may be placed on the other so that the equal angles coincide. Let them be so placed, as in Fig. 148, where ABC, ADE are the triangles and $\angle A$ is the same in both. The theorem then states that

$$\frac{\text{area } \triangle ABC}{\text{area } \triangle ADE} = \frac{\overline{AB} \times \overline{AC}}{\overline{AD} \times \overline{AE}}$$

and this is to be proved.

Join BE and draw $EF \perp AB$, and $BG \perp AC$ produced. Then triangles ABC, ABE have the same altitude BG and bases AC, AE. According to IX (2) above, therefore,

$$\frac{\text{area } \triangle ABC}{\text{area } \triangle ABE} = \frac{AC}{AE}.$$

Also triangles ABE, ADE have the same altitude EF and bases AB, AD. According to IX (1) therefore,

$$\frac{\text{area } \triangle ABE}{\text{area } \triangle ADE} = \frac{AB}{AD}.$$

Multiplying the first ratios, and also the second ratios, of these proportions, and cancelling area $\triangle ABE$ in the first product,

$$\frac{\text{area } \triangle ABC}{\text{area } \triangle ADE} = \frac{\overline{AB} \times \overline{AC}}{\overline{AD} \times \overline{AE}} \qquad \text{—Q.E.D.}$$

The second of the area proportion theorems is:

XIV. *The areas of two similar triangles are in the same ratio as the squares of any two corresponding sides.*

In Fig. 149 let (a) and (b) be two similar triangles, with AB and DE as two corresponding sides. We then have to show that

$$\frac{\text{area } \triangle ABC}{\text{area } \triangle DEF} = \frac{\overline{AB}^2}{\overline{DE}^2}.$$

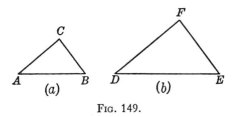

Fig. 149.

Since the triangles are similar, then by definition their corresponding angles are equal; hence $\angle A = \angle D$. According to XIII above, therefore,

$$\frac{\text{area } \triangle ABC}{\text{area } \triangle DEF} = \frac{\overline{AB} \times \overline{AC}}{\overline{DE} \times \overline{DF}}.$$

But since the triangles are similar their sides are proportional; hence, $\frac{AC}{DF} = \frac{AB}{DE}$. Substituting this equal value of $\frac{AC}{DF}$ in the preceding proportion, we have

$$\frac{\text{area } \triangle ABC}{\text{area } \triangle DEF} = \frac{\overline{AB}^2}{\overline{DE}^2}. \qquad \text{—Q.E.D.}$$

We give next a demonstration of the *Pythagorean Theorem* which is of very great historical interest.

XV. *The square on the hypotenuse of a right triangle equals the sum of the squares on the legs.*

In Fig. 150 let ABC be a right triangle having the right $\angle BAC$, with the square $BCED$ constructed on the hypotenuse BC as a side and the squares *ABFG*, *ACKH* on the legs *AB*, *AC* respectively. Then the theorem states and we are to prove that $\square BE = \square BG + \square CH$, where the expression $\square BE$ means "area of square *BCED*," etc.

Draw *AL* ∥ *BD* or *CE*, and join *AD*, *FC*. Then because ∠*BAC* is a right angle (by definition of a rt. △), the two straight lines *AC*, *AG* make with *AB* two right angles; hence *CAG* is a straight angle and *CG* a straight line (def. st. angle, equals 2 rt. angles), and for the same reasons *BH* is also a straight line.

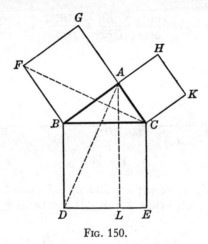

Fig. 150.

Now ∠*DBC* = ∠*FBA*, as both are right angles. Add to each of these angles the angle *ABC*; then the whole ∠*DBA* = ∠*FBC*; also *AB* = *BF* and *DB* = *BC*, as sides of the same square. Therefore triangles *ABD*, *FBC*, having two sides and included angle equal are congruent (IV, 39) and therefore also equal.

Next, the rectangle *BL* equals twice the triangle *ABD* because they have the same base *BD* and the same altitude *DL* (between the same parallels *BD* and *AL*) according to VI, 72, and VII above. And for the same reasons square *BG* = 2 × (∠*FBC*). But by axiom 5, article 20, doubles of equals are equal; hence $\square BL = \square BG$.

In the same manner, by joining *AE*, *BK* it is shown that $\square CL = \square CH$. Hence, adding, $\square BL + \square CL = \square BG + \square CH$. But $\square BL + \square CL = \square BE$. Therefore, $\square BE = \square BG + \square CH$.

—Q.E.D.

This is the proof of the theorem of Pythagoras which was given by Euclid in his *Elements of Geometry* (article 4). Euclid also showed that the theorem is true for any *similar* figures, as well as squares, constructed on the sides of the right triangle, such as semi-circles, equilateral triangles, parallelograms, etc.

75. Note on the Pythagorean Theorem. Something of the history of Pythagoras, in whose honor this theorem is named, is sketched in article 3. While he was the first to prove that the proposition is true of all right triangles, the ancient Egyptians had known for a long time that it is true for certain particular triangles. Thus, they knew that a triangle whose sides are 3, 4 and 5 units in length has a right angle included between the shorter sides, 3 and 4, opposite the longest side 5, and that the sum of the squares $3^2 + 4^2$ equals 5^2, the square of the longest.

From very early times the Egyptians utilized this fact to determine the true east-west direction. Thus they tied a rope 12 units long into a loop and stretched it around three points 5, 4 and 3 units apart at the corners of a triangle. The two shorter sections then made a right angle and when one of these was set in the north-south direction, as determined by means of the pole star, the other was in the east-west line. Other right triangles such as those of sides 5, 12, 13 or 6, 8, 10, etc., were used for the same purpose, and also to lay off right angles in surveying, building and the like, and those who were experts in selecting and laying out such triangles made it a regular business and were known as "rope-stretchers." This method is still used by carpenters and others for laying out right angles, but instead of a single rope loop it is more usual and convenient to use a yard stick (3 feet) and two other sticks of 4 and 5 foot lengths, or else a two-yard (6 feet), an 8- and a 10-foot pole, laid on the ground to form the right triangle.

Since the time of Pythagoras many other proofs of the theorem have been given, and it has been found to be one of the fundamental theorems of modern geometry in all its branches and also of trigonometry. The proof we have given in article 64 (XIX) is thought by historians to be one of the first given by Pythagoras himself, while the one given above (XV) is, as we have stated, the statement and proof given by Euclid. It is to be noted that in the form of XIX, 64, the theorem refers to the *algebraic* square of the numbers representing the *lengths* of the sides of the right triangle, while the form given by Euclid refers to the *areas* of the *geometrical* squares constructed with sides equal to those of the triangle. The two are of course equivalent, for the area of a square in square units is the product of the length of a side by itself, that is, the algebraic square of the length (V, 74).

Pythagoras seems to have discovered more than one method of proving his theorem. One other thought to have been given by him is given in the ARITHMETIC of this series. A proof devised by the noted Hindu mathematician Bhaskara (article 6) is also given in the TRIGONOMETRY of this series. Collections of many other proofs of the theorem have been published, and among these is included a demonstration given by a President of the United States, James A. Garfield, who was also an army officer, a mathematician, and a minister of one of the branches of the Protestant Christian Church.

76. Areas of Trapezoids and Polygons. As shown in Fig. 151 (*b*) a trapezoid, as *ABCDE*, can be considered as formed of two triangles,

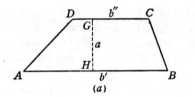

FIG. 151.

ABD and *BCD*, whose bases *AB*, *CD* are the parallel lower and upper bases of the trapezoid, and whose altitudes *DE*, *BF* (perpendicular to the bases) are both the same as $GH = a$ of the trapezoid.

The areas of the triangles in (*b*) are, according to IV, 74,

$$\text{area} \triangle ABD = \tfrac{1}{2}(\overline{DE} \times \overline{AB}) = \tfrac{1}{2}ab'$$
$$\text{area} \triangle BCD = \tfrac{1}{2}(\overline{BF} \times \overline{CD}) = \tfrac{1}{2}ab''$$

Adding, $\quad \text{area} \triangle ABD + \text{area} \triangle BCD = \tfrac{1}{2}ab' + \tfrac{1}{2}ab''.$

But the sum of the two triangle areas is the area of the trapezoid, and

$$\tfrac{1}{2}ab' + \tfrac{1}{2}ab'' = \tfrac{1}{2}a \times (b' + b'') = \frac{a}{2}(b' + b''). \quad \text{Hence,}$$

$$\text{area trap. } ABCD = \frac{a}{2}(b' + b'').$$

In words this is, for any trapezoid,

ART. 76 DIMENSIONS AND AREAS OF PLANE FIGURES   175

XVI. *The area of a trapezoid is half the product of the altitude and the sum of the bases.*

The altitude and bases of a trapezoid are its dimensions.

In case either diagonal and the slant heights of a trapezoid are known, as *AD*, *BD*, *BC*, Fig. 151, the area can be found by considering these as the sides of two triangles, as *ABD*, *BCD*, and applying X, 76, to each triangle. The sum of the two areas is then the area of the trapezoid.

We have seen how the area of a rectangle, a parallelogram or a trapezoid can be found by dividing it into two triangles by drawing a diagonal, the triangles being equal in the case of the rectangle and the parallelogram but not in the trapezoid. Similarly, any quadrilateral may be divided into two triangles by a diagonal and the areas of each found separately by applying the rules VII or X in article 74.

Similarly the area of any polygon may be found by dividing it into triangles by drawing all the diagonals from any one vertex, as in Fig. 152(*a*).

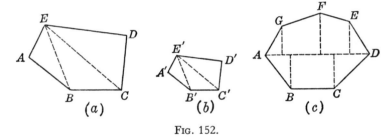

Fig. 152.

The area of an irregular polygon may also be found by drawing the longest diagonal, as *AD*, Fig. 152(*c*), and drawing perpendiculars to this diagonal from each of the other vertices. In this way, the polygon is divided into triangles, rectangles, trapezoids or other quadrilaterals.

In either case (*a*) or (*c*), the area of the polygon is the sum of the areas of the separate figures into which it is divided.

By the method of division of polygons into triangles an interesting relation between similar polygons can be brought out. It is

XVII. *The areas of two similar polygons are in the same ratio as the squares of any two corresponding sides.*

Let *A*, *A'* represent the areas of the similar polygons (*a*), (*b*), Fig.

152. Then, the relation stated in the theorem is $A:A' = \overline{AB}^2 : \overline{A'B'}^2$, which is easily proved.

For, according to XIV, 76, $\dfrac{\overline{AB}^2}{\overline{A'B'}^2} = \dfrac{\triangle ABE}{\triangle A'B'E'} = \dfrac{\overline{BE}^2}{\overline{B'E'}^2} = \dfrac{\triangle BCE}{\triangle B'C'E'} =$, etc. That is,

$$\dfrac{\triangle ABE}{\triangle A'B'E'} = \dfrac{\triangle BCE}{\triangle B'C'E'} = \dfrac{\triangle CDE}{\triangle C'D'E'}.$$

This is a continued proportion, a series of equal ratios; therefore,

$$\dfrac{\triangle ABE + \triangle BCE + \triangle CDE}{\triangle A'B'E' + \triangle B'C'E' + \triangle C'D'E'} = \dfrac{\triangle ABE}{\triangle A'B'E'},$$

according to VIII, 61. But the sum of the triangles in the numerator of the fraction on the left is the area A of the polygon $ABCDE$, and similarly the sum in the denominator is A'. Also, as found above $\dfrac{\triangle ABE}{\triangle A'B'E'} = \dfrac{\overline{AB}^2}{\overline{A'B'}^2}$. Therefore

$$\dfrac{A}{A'} = \dfrac{\overline{AB}^2}{\overline{A'B'}^2}. \qquad \text{—Q.E.D.}$$

We include here an important result which pertains to both triangles and circles. It is

XVIII. *The area of any triangle which is inscribed in a circle is the product of its three sides divided by twice the diameter of the circle.*

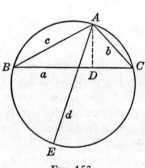

Fig. 153.

In Fig. 153 let ABE be the given circle, of diameter $AE = d$, and $\triangle ABC$ any inscribed triangle, having sides $BC = a$, $AC = b$, $AB = c$ and altitude $AD = h$ on side BC.

Then according to XXIV, 65, $bc = dh$. Multiply this equation by a; then $abc = dah$. But the area of $\triangle ABC$ is $A = \tfrac{1}{2}ah$; hence, $ah = 2A$ and the last equation can be written $abc = 2Ad$. Solving this for A,

$$A = \dfrac{abc}{2d}.$$

The area of the inscribed triangle is therefore the *product of the sides divided by twice the diameter.*

ART. 77 DIMENSIONS AND AREAS OF PLANE FIGURES

77. Illustrative Examples. In the present chapter, and similarly in the chapters which follow, we deal more with the measurement of geometrical figures than with the relations between their geometrical properties as covered in the preceding chapters. On this account the illustrative examples now to be worked out are more of the nature of numerical problems than of explanations or descriptions of principles and methods of measurement. They will thus depend more on calculation than on drawing.

Example 1. A town lot has the shape and dimensions shown in Fig. 154. Find its area in square feet and the value at $1500 per acre.

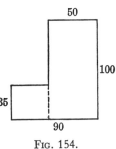

Fig. 154.

Solution. Considered as a whole the figure of the lot is not that of any of the simple geometric figures we have studied, but if it be divided as shown by the dotted line it is seen to consist of two rectangles.

The larger of these rectangles is 50 × 100 ft. and its area is therefore 50 × 100 = 5000 sq. ft. The width of the smaller rectangle is 35 ft. and its length is 90 − 50 = 40 ft. The area is therefore 35 × 40 = 1400 sq. ft. The total area is therefore 5000 + 1400 = 6400 sq. ft.

Since 1 acre = 43,560 sq. ft., the area of the lot in acres is 6400 ÷ 43,560 = 0.147 acre. The value at $1500 per acre is then .147 × 1500 = $220.50.

Example 2. What must be the length of the side of a flat square box which has an area of $1\frac{13}{36}$ sq. ft.?

Solution. The area of a square of side a is $A = a^2$. Therefore the side is found from the area as $a = \sqrt{A}$. Expressed in square inches the area of our given square is $A = 144 \times 1\frac{13}{36} = 196$ sq. in. The length of the side is therefore $a = \sqrt{196} = 14$ inches.

Example 3. Find the area of the two gable ends of a house which is 30 ft. wide with the ridge of the roof 20 ft. above the center of the cross beams on which the rafters rest.

Solution. The shape of the gable ends is triangular, with the width of the building as base, $b = 30$ ft., and the height of the ridge as altitude, $a = 20$ ft. The area of each gable end is therefore $A = \frac{1}{2}ab$, and hence for the two it is $ab = 20 \times 30 = 600$ sq. ft.

Example 4. Suppose the dimensions of the gables in the preceding example to be given by stating that the width is 30 and the length of the rafters is 25 ft. How is the area found in this case?

Solution. This description shows the triangular gables to be isoceles with legs $a = 25$ and base $c = 30$ ft. According to XI (2), 76, therefore, the area of each is $\frac{c}{4}\sqrt{4a^2 - c^2}$, and the total area is twice this amount, or $\frac{c}{2}\sqrt{4a^2 - c^2} = \frac{30}{2}\sqrt{4(25)^2 - (30)^2} = 15\sqrt{4 \times 625 - 900}$ $= 15\sqrt{1600} = 15 \times 40 = 600$ sq. ft., as before.

Example 5. A "star" or "rosette" cut from cardboard is in the form of a 2-inch square with an equilateral triangle on each side. How much silvered paper is required to cover both sides of the star?

Solution. The area of the square is $2^2 = 4$ sq. in., and according to XI (2), 76, the area of each equilateral triangle is $\frac{a^2}{4}\sqrt{3}$, where a is the side, which is here the same as that of the square.

The area of the *four* triangles is therefore $a^2\sqrt{3} = 2^2\sqrt{3} = 4\sqrt{3} = 6.928$ sq. in. The total area of the figure is therefore $4 + 6.928 = 10.928$ sq. in., and to cover both sides will require $2 \times 10.928 = 21.856$ or very nearly 22 sq. in. of paper.

Example 6. A parallelogram has a base of 24 inches and a *slant* height of 12 inches, which is inclined at an angle of 30° to the base. What is the area of the parallelogram?

Solution. According to VI, 72, the area is the product of the base and *altitude*, not slant height. This means that in Fig. 145(a) we have $AB = 24$ in. and to find the area we need to know DE.

What we have given, however, is the *slant* height (or width) $AD = 12$ in. and $\angle A = 30°$. But AD and the altitude DE form with the base the right $\triangle AED$, and according to article 41, in this right triangle having a 30° angle the hypotenuse is equal to twice the length of the shorter leg. The shorter leg, which is, in this case, the required altitude, opposite the 30° angle, is therefore equal to half the hypotenuse AD. Therefore $DE = 6$ in., and the area is equal to $AB \times DE = 24 \times 6 = 144$ sq. in. $= 1$ sq. ft.

Example 7. Two parallel streets are 264 ft. apart, and are crossed by two others which are not parallel; the block enclosed by the four is therefore not rectangular nor is it a parallelogram. If this block is 300 ft. long on one side and 250 ft. on the other, what is its area?

Solution. The shape of the block is trapezoidal, with one of the

ART. 77 DIMENSIONS AND AREAS OF PLANE FIGURES   179

parallel sides 300 and the other 250 ft. The bases of the trapezoid are, therefore, $b' = 300$, $b'' = 250$, and the altitude $a = 264$ ft. (see Fig. 151(a)).

According to XVI, 76, therefore, the area of the block is

$$A = \tfrac{264}{2}(300 + 250) = 132 \times 550 = 72{,}600 \text{ sq. ft.}$$

Example 8. In Fig. 152(a) suppose the sides of the polygon $ABCDE$ are $AB = 3$, $BC = 3$, $CD = 3$, $DE = 5$, $EA = 2$ inches, and the diagonals $BE = 4$, $EC = 6$ inches. Find the area of the polygon.

Solution. The diagonals divide the polygon into the three triangles ABE, EBC, CDE, and as listed the sides of these are:

$\triangle ABE$	$\triangle EBC$	$\triangle CDE$
$AB = 3$	$EB = 4$	$CD = 3$
$BE = 4$	$BC = 3$	$DE = 5$
$EA = 2$	$CE = 6$	$EC = 6$

The three sides of each triangle being given the area of each can be found by the formula of X, 76, and the area of the polygon is the sum of the areas of the triangles.

For $\triangle ABE$, calling the sides in the order given a, b, c, we have $s = \tfrac{1}{2}(3 + 4 + 2) = 4.5$, $s - a = 4.5 - 3 = 1.5$; $s - b = 4.5 - 4 = 0.5$; $s - c = 4.5 - 2 = 2.5$; and hence the area is

$$\sqrt{4.5 \times 1.5 \times .5 \times 2.5} = \sqrt{8.4375} = 2.91 \text{ sq. in.}$$

For $\triangle EBC$, $s = \tfrac{1}{2}(4 + 3 + 6) = 6.5$; $s - a = 2.5$; $s - b = 3.5$; $s - c = 0.5$; and the area is $\sqrt{6.5 \times 2.5 \times 3.5 \times .5} = 5.33$ sq. in.

For $\triangle CDE$, $s = \tfrac{1}{2}(3 + 5 + 6) = 7$; $s - a = 4$; $s - b = 2$; $s - c = 1$; and the area is $\sqrt{7 \times 4 \times 2 \times 1} = 7.48$ sq. in.

The area of the polygon is therefore $A = 4.11 + 7.55 + 7.48 = 19.14$ sq. in.

Example 9. A fire ladder is set up in the street 20 ft. from the wall of a building. How long must the ladder be to reach a fifth story window 50 ft. above the street level?

Solution. As the street level is horizontal and the wall vertical they are perpendicular, and the inclined ladder forms with the ground and wall a right triangle, of which the ladder is the hypotenuse.

According to the *Pythagorean Theorem* therefore, the length of the ladder is $L = \sqrt{(20)^2 + (50)^2} = \sqrt{400 + 2500} = \sqrt{2900} = 53.85$ ft. A 55-foot ladder will therefore reach the window.

Let us consider the question: How far from the wall must a 60-foot ladder stand in order that it may just reach the same window? Here the 60-foot ladder is the hypotenuse of the right triangle, the height to the window, 50 feet, is one leg, and the distance from the wall to the foot of the ladder is the other (unknown) leg.

Therefore $(60)^2 = (50)^2 + (\text{Distance})^2$, or $(\text{Distance})^2 = (60)^2 - (50)^2 = 3600 - 2500 = 1100$, and the Distance $= \sqrt{1100} = 33.166$ ft. or about 33 ft. 2 in.

As it is not the object of these examples to show the applications of the principles of this chapter in the arts, sciences and industry, but simply to illustrate their meaning and their use in calculation, the illustrations given are considered sufficient.

78. Exercises and Problems.

1. How many tiles, each 8 inches square, will be required to pave a rectangular space 18 × 30 feet?

2. Compare the areas of two rectangles whose altitudes are equal and whose bases are 10 and 6 inches.

3. Draw a rectangle 5 × 6 inches and draw both its diagonals. Show that either diagonal divides the rectangle into two congruent triangles and find the area of each.

4. In a certain parallelogram the *sides* are 10 and 14 inches, and the acute angle is 30 degrees. Find the altitude and the area.

5. Draw several triangles having the same base and altitude but different shapes. Are the triangles congruent? Are they equal?

6. What is the area of a triangle of any shape whose base is 12 and altitude 5 inches?

7. What are the area and hypotenuse of a right triangle whose legs are 5 and 12 inches?

8. What are the area and side of a square whose side is the hypotenuse of a right triangle with legs of 7 and 14 inches?

9. Show that the area of an isosceles right triangle is one fourth of that of the square having the hypotenuse as side.

10. Find the area of a trapezoid whose bases are 9 and 15 inches and whose altitude is 12 inches.

11. A long narrow tract of land between two parallel straight roads fronts 220 yards on one road and 180 yards on the other. The two end boundaries are straight fences, one of which is 50 yards long and makes an angle of 60 degrees with the road on the longest front. How many acres does the tract contain?

12. The three sides of a triangle are 5, 6, 7 inches. Find the area of the triangle and the diameter of the circumscribed circle.

13. A four-legged table is braced by four bars across the sides and ends and by four tight wires stretched from the top of each leg to the bottom of the

diagonally opposite leg. If the table is 3 ft. high and the end and side bars are 4 and 6 ft. long, how long is each of the wires?

14. A baseball diamond is a square 30 yards on a side. How far is the home plate from second base?

15. A tree is broken 30 feet above the level ground. The top strikes the ground 40 feet from the foot, while the other end of the broken part remains attached to the stump. How high was the tree?

Chapter 8

REGULAR POLYGONS AND MEASUREMENT OF THE CIRCLE

79. Regular Polygons and Circles. In article 49 a *polygon* is defined as a closed plane figure of any number of sides formed by straight lines. Polygons are named according to the number of their sides or angles, usually the sides. A list of a few polygons is given in article 49, containing the names and corresponding numbers of sides.

The sides of a polygon may have any lengths whatever but the values of the angles are restricted in accordance with the relation stated in XL, 50. Provided the sum of the interior angles is not greater than $(n - 2) \times 180°$, however, where n is the number of sides, the individual angles may have any values.

If all the sides of a polygon are of equal length the polygon is said to be *equilateral*. If all the angles of a polygon are equal it is said to be *equiangular*.

If a polygon is *both* equilateral and equiangular it is called a *regular* polygon. A regular polygon may be of any size, but all regular polygons of the same number of sides are of the same shape, that is, they are similar.

The value of each of the angles of a regular polygon depends on the number of equal sides and is found from the relation given in XLII, 50, namely, $\left(1 - \dfrac{2}{n}\right) \times 180°$.

The sum of the exterior angles (formed by one side being extended through each vertex, all in the same sense as we proceed around the polygon) of any polygon is 360° (XLIII, 50) and hence each exterior angle of a *regular* polygon of n sides is $\dfrac{360°}{n}$.

The regular polygon of three sides is the equilateral triangle and that of four sides is the square. Those of other numbers of sides are the same as those given in the list in article 49, with the description "regular" added to each. This list is given here for convenience of reference.

Number of Sides	Name of Regular Polygon
3	equilateral triangle
4	square
5	regular pentagon
6	hexagon
7	heptagon
8	octagon
9	nonagon
10	decagon
11	undecagon
12	dodecagon
15	pentadecagon

If a regular polygon has each of its vertices on a circle it is said to be *inscribed* in the circle, and the circle, passing through the vertices, is said to be *circumscribed* about the polygon.

If a regular polygon has all its sides tangent to a circle it is said to be circumscribed about the circle, and the circle is inscribed in the polygon.

The center of a circle circumscribed about a regular polygon is called the *center* of the polygon, and the radius of this circle, which is a line from the center to a vertex of the polygon, is also called the *radius* of the regular polygon.

The *perpendicular* from the center of a regular polygon to a side is called the *apothem* of the polygon.

As already defined in article 49, the sum of the lengths of the sides of a polygon is called the *perimeter* of the polygon. This term is also applied to the lines forming the polygon without reference to their lengths.

There are many very interesting relations between circles and inscribed and circumscribed regular polygons, and these provide the methods of deriving the rules for the measurement of the circle.

We proceed to the development of some of these relations.

80. Relations of Regular Polygons and Circles. The first step in the development of the relations between regular polygons and circles is to show that

I. *An equilateral polygon inscribed in a circle is a regular polygon.*

In order that an equilateral polygon be regular it must also be equiangular. To prove the theorem, therefore, it is only necessary to show that the inscribed equilateral polygon has equal angles.

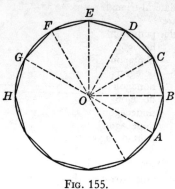

FIG. 155.

In Fig. 155 let $ABCD\ldots$ be the equilateral polygon inscribed in the circle O. We have to show that the angles at A, B, C, etc., are equal.

Draw the radii OA, OB, OC, etc., to the vertices, making all the triangles AOB, BOC, COD, etc., isosceles. Then since the chords AB, BC, CD, etc., are all equal, their subtended central angles at O, formed by the radii, are all equal (III, 54). Therefore the isosceles triangles are all congruent (two sides and including angle equal) and hence their base angles are all equal. Therefore the sum of the two base angles of the triangles at each vertex is the same for every vertex. That is, $\angle ABC = \angle BCD = \angle CDE =$ etc., and the polygon is equiangular, and therefore regular, as it was given equilateral.

The question next arises: Can a circle be inscribed in or circumscribed about every regular polygon? This is, as we shall show, answered by the statement that

II. *A circle may be circumscribed about or inscribed in any regular polygon.*

In Fig. 156 let $ABCDE$ be a regular polygon, and describe the circle through any three vertices, say A, B, C (VII, 54). Let O be the center of this circle and draw OA, OB, OC, OD. Then, since the polygon is regular,
$$\angle ABC = \angle BCD.$$
Also, in isosceles $\triangle OBC$, $\angle OBC = \angle OCB$.
Subtracting, $\triangle ABC - \angle OBC = \angle BCD - \angle OCB$,
or $\angle OBA = \angle OCD.$

Therefore, since also $OB = OC$ as radii of the same circle, and $AB = CD$ as sides of the regular polygon, triangles OBA, OCD are congruent (IV, 39), and hence the corresponding sides $OA = OD$. Therefore OD is also a radius of the circle, or, the circle through A, B, C also passes through D.

In the same manner it is shown that the circle also passes through E, etc., and hence is circumscribed about the polygon.

Now the sides of the regular polygon are equal and are chords of the circumscribed circle. They are therefore at the same distance OF from the center (VI, 54), and hence a circle with center O and radius OF will touch each side AB, BC, CD, etc. This circle is therefore inscribed in the polygon. (This radius is the *apothem* of the polygon.)

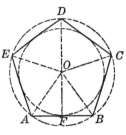

Fig. 156.

Both parts of the theorem are thus proved.

III. *Two regular polygons of the same number of sides are similar.*

Let each regular polygon have n sides. Then each angle of each polygon is $\left(1 - \dfrac{2}{n}\right) \times 180°$ and hence the same in both polygons.

Let AB, BC, CD, etc. be the sides of one polygon, and $A'B'$, $B'C'$, $C'D'$, etc. those of the other. Then
$$AB = BC, \text{ etc.}$$
and
$$A'B' = B'C', \text{ etc.}$$
Dividing these equalities (equality division axiom) we have
$$\frac{AB}{A'B'} = \frac{BC}{B'C'}, \text{ etc.}$$

Thus, the sides of the polygons are proportional, and since the angles are also equal, the polygons are therefore similar.

IV. *The perimeters of two regular polygons of the same number of sides are to each other as their radii and also as their apothems.*

In Fig. 157 let (*a*) and (*b*) represent the polygons with centers O, O' radii OA, $O'A'$ apothems OM, $O'M'$ and perimeters P, P'. The theorem then states that $P:P' = OA:O'A' = OM:O'M'$.

Since the polygons are similar, then $\dfrac{AB}{A'B'} = \dfrac{BC}{B'C'} = \dfrac{CD}{C'D'} =$ etc.

Therefore $\dfrac{AB + BC + CD + \text{etc.}}{A'B' + B'C' + C'D' + \text{etc.}} = \dfrac{AB}{A'B'}$, by VIII, 61. But $AB + BC + CD + \text{etc.} = P$ and $A'B' + B'C' + C'D' + \text{etc.} = P'$.

Therefore, $\dfrac{P}{P'} = \dfrac{AB}{A'B'}$.

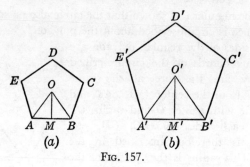

Fig. 157.

As in the proof of I above $\angle AOB = \angle A'O'B'$, and hence, as triangles AOB, $S'O'B'$ are isosceles, they are also similar. Hence, $\dfrac{AB}{A'B'} = \dfrac{OA}{O'A'}$, and also in the similar right triangles OMA, $O'M'A'$, $\dfrac{AB}{A'B'} = \dfrac{OM}{O'M'}$. Therefore, by the equality axiom,

$$\frac{P}{P'} = \frac{OA}{O'A'} = \frac{OM}{O'M'}. \qquad \text{—Q.E.D.}$$

As the proposition XVII, 76, applies to any polygon, we have also,

V. *The areas of two regular polygons of the same number of sides are to each other as the squares of their sides, radii or apothems.*

The two following propositions are extremely important.

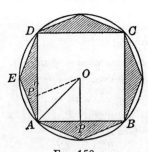

Fig. 158.

VI. *If the number of sides of a regular inscribed polygon is indefinitely increased the apothem approaches the radius as a limit.*

In Fig. 158 let OA be the radius, AB the side, and OP the apothem of a circle and its inscribed square $ABCD$. Then the apothem OP is less than the radius OA.

Now suppose the regular polygon of twice the number of sides (octagon) to be inscribed by *doubling the sides*. A side of the new polygon, as AE, being shorter than the original side, is farther from the center and nearer to the circle, and hence the apothem OP' of the new polygon is longer than OP. It is still shorter than OA, however.

Suppose now that the number of sides is again doubled, and this process repeated, and continued indefinitely. Then the apothem of each new polygon is longer than that of the preceding one and ultimately may be made as nearly equal to the radius as desired. That is, the apothem approaches the radius as a limit (article 55).

> VII. *If the number of sides of a regular inscribed polygon is indefinitely increased the perimeter and area respectively approach the circumference and area of the circle as limits.*

In Fig. 158 the area of the square is considerably less than that of the circle, and its perimeter is considerably less than the circumference of the circle. The area of the octagon, however, is more than that of the square, by the amount of the shaded area, and hence more nearly equal to that of the circle. Similarly the perimeter of the octagon is nearer to the circumference than that of the square.

If now the regular polygon of 16 sides is inscribed its perimeter and area will be still more nearly equal to the circumference and area of the circle, and if the number of sides is doubled again, and this process is continued indefinitely, the perimeter and area of the polygon become more and more nearly equal to the circumference and area of the circle, and can ultimately be made to differ from those by as small amounts as we please.

That is, the perimeter and area of the polygon approach the circumference and area of the circle as limits (article 55).

Formal algebraic-geometric proofs of these two theorems, as well as of XV, 56, and I, 72 are given in advanced treatises, based on the detailed theory of limits, but the explanations given here and illustrated in the figure (Fig. 158) are simpler and convince us of the truth of the propositions.

These results will be used in developing some of the chief properties of the circle, particularly in connection with the rules for its measurement.

> VIII. *The area of a regular polygon equals half the product of its perimeter and apothem.*

In Fig. 159 let $ABCD\ldots$ be a regular polygon with center O and apothem $OM = a$, and let its perimeter and area be P, A respectively.

Draw the radii OA, OB, OC, etc., forming the equal triangles AOB, BOC, COD, etc. Then the base of each of these triangles, as $\triangle AOB$, is a side (AB) of the polygon and the altitude (OM) is the apothem, which is the same for all. Therefore the area of

$$\triangle AOB = \tfrac{1}{2}\overline{OM} \times \overline{AB}.$$

Similarly
$$\triangle BOC = \tfrac{1}{2}\overline{OM} \times \overline{BC},$$
$$\triangle COD = \tfrac{1}{2}\overline{OM} \times \overline{CD},\text{ etc.}$$

Adding,
$$\triangle AOB + \triangle BOC + \triangle COD + \ldots$$
$$= \tfrac{1}{2}\overline{OM} \times \overline{AB} + \tfrac{1}{2}\overline{OM} \times \overline{BC} + \tfrac{1}{2}\overline{OM} \times \overline{CD} + \ldots$$
$$= \tfrac{1}{2}\overline{OM} \times (\overline{AB} + \overline{BC} + \overline{CD} + \ldots)$$

That is,

(Sum of $\triangle$) = $\tfrac{1}{2}\overline{OM} \times$ (Sum of sides of polygon).

But the sum of the triangle areas is the area of the polygon, and the sum of the sides of the polygon is its perimeter. Therefore,

Area of Polygon = $\tfrac{1}{2}\overline{OM} \times$ (Perimeter),

or,
$$A = \tfrac{1}{2}a \times P = \tfrac{1}{2}aP,$$

as was to be proved.

Since $\triangle AMO$ (Fig. 159) is a right triangle and $\angle OAM$ or $\angle AOM$ is known when the number of sides is known, then as soon as the length of the side AB (and hence AM, its half) is known the apothem $OM = a$ can be calculated for any regular polygon from the properties of right triangles. Therefore the area of any regular polygon can be calculated when the number and length of its sides are given. Formulas have been developed for this calculation and these are given in the following table for a number of regular polygons. In these formulas s is the length of the side.

Regular Polygons		
No. Sides	Name	Area
3	Equil. triangle	$0.433s^2$
4	Square	$1.000s^2$
5	Pentagon	$1.720s^2$
6	Hexagon	$2.598s^2$
7	Heptagon	$3.634s^2$
8	Octagon	$4.828s^2$
9	Nonagon	$6.182s^2$
10	Decagon	$7.694s^2$
11	Undecagon	$9.366s^2$
12	Dodecagon	$11.196s^2$
15	Pentadecagon	$17.642s^2$

IX. The side of a regular inscribed hexagon equals the radius.

In Fig. 160 let $ABCDEF$ be a regular inscribed hexagon with center O and radius OA. We have to show that $AB = OA$.

Draw the radii OB, OC, etc. to each vertex, forming the triangles AOB, BOC, COD, etc., and the six central angles at O. Then the chords AB, BC, CD, etc., are equal to each other and hence the six central angles at O are equal (III, 54). Therefore each central angle is $\frac{360°}{6} = 60$ degrees.

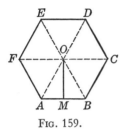

Fig. 159.

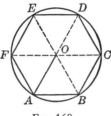

Fig. 160.

The sum of the other two angles in each triangle is then 120° (total sum equals 180° in each triangle), and as the triangles are isosceles since the legs are the equal radii, these other two angles in each triangle are equal. Therefore each is half of 120°, or 60 degrees. Thus each of the three angles of each triangle is 60 degrees and hence the triangles are equiangular and equilateral, and hence $AB = OA$, as was to be proved.

81. Inscription of Regular Polygons. The problem of inscribing regular polygons in circles is one of the most interesting in all geometry and in art and architecture it is of very great importance. It also provides application of many of the geometrical principles we have already discovered and leads directly to the very practical matter of the measurement of the circle.

We give now the solutions of several of the construction problems of inscription. These cannot be given strictly in the order of increasing numbers of sides, because the construction of a polygon of a smaller number of sides sometimes depends on that of the larger number of sides.

X. To inscribe a square in a given circle.

In Fig. 161 let O be the center of the circle and through O draw two perpendicular diameters AC, BD. Join AB, BC, CD, DA. Then ABCD is the required square.

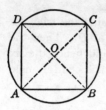

FIG. 161.

For, each ∠ABC, BCD, etc., is inscribed in a semi-circle, and hence is a right angle (XVIII, 54). Also each central angle at O is a right angle (AC ⊥ BD), and hence the chords AB, BC, etc., are all equal (II, III, 54). Therefore, since the four sides are equal and the four angles are right angles, the quadrilateral ABCD is a square, and since its vertices are on the circle it is inscribed.

XI. *By doubling the number of sides of the inscribed square, the inscribed regular octagon is obtained; and hence also the regular inscribed polygons of* 16, 32, 64, *etc., sides.*

XII. *To inscribe a regular hexagon in a given circle.*

The problem is solved at once by referring to proposition IX in the preceding article. Thus apply the radius to the circumference as a chord, six times in succession, as AB, BC, CD, etc., in Fig. 160.

XIII. *To inscribe an equilateral triangle in a given circle.*

Inscribe the regular hexagon and join the alternate vertices, as A, C, E, Fig. 160.

XIV. *By doubling the number of sides of the regular inscribed hexagon the regular inscribed dodecagon is obtained; and hence also the regular inscribed polygons of* 24, 48, 96, *etc., sides.*

XV. *To inscribe a regular pentagon in a given circle.*

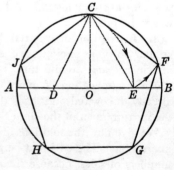

FIG. 162.

Let Fig. 162 represent the given circle, with center O. Draw a diameter AB and at O erect OC ⊥ AB. Bisect the radius at D and with center D and radius DC lay off DE = DC. Then with center C and radius CE draw the arc EF, intersecting the circle at F, and draw the chord CF. CF is then a side of the required regular pentagon, and by applying CF five times as a chord, as at CF, FG, GH, HJ, VC, the pentagon is obtained.

The proof of this construction requires a knowledge of trigonometry and will not be given here.

XVI. *By doubling the number of sides of the inscribed regular pentagon the inscribed regular decagon is obtained; and hence also the regular polygons of* 20, 40, 80, *etc., sides.*

It is also possible to inscribe the regular decagon (10 sides) directly without first inscribing the regular pentagon, by means of a special proportion called "extreme and mean proportion," which is not discussed in this book. The regular pentagon may then be obtained by joining alternate vertices of the decagon. The regular polygon of 20 sides is sometimes called the *icosagon*.

82. Lengths of Sides of Inscribed Regular Polygons. In Fig. 163 let ABC be the equilateral triangle inscribed in a circle of radius $OC = R$, and let S_3 represent the length of the side. Draw the altitude CD. Then, since $\triangle ABC$ is also isosceles, D is the midpoint of AB and hence $AD = \tfrac{1}{2}S_3$ (X, 39).

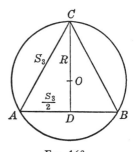

Fig. 163.

The altitude CD being thus the perpendicular bisector of chord AB passes through the center O (IV, 54), which is also the center of the triangle, and according to XVII, 42, $OC = R = \tfrac{2}{3}\overline{CD}$, or $\overline{CD} = \tfrac{3}{2}R = 1\tfrac{1}{2}R$.

Thus in the right $\triangle ADC$ the sides are $AC = S_3$, $AD = \dfrac{S_3}{2}$, $CD = \dfrac{3R}{2}$, and by the *Pythagorean Theorem* $\overline{AC}^2 = \overline{AD}^2 + \overline{CD}^2$. That is,

$$S_3{}^2 = \left(\frac{S_3}{2}\right)^2 + \left(\frac{3R}{2}\right)^2 = \frac{S_3{}^2}{4} + \frac{9R^2}{4}.$$

Clear fractions,
$$4S_3{}^2 = S_3{}^2 + 9R^2.$$

Transpose,
$$3S_3{}^2 = 9R^2, \quad \text{or} \quad S_3{}^2 = 3R^2.$$

$$\therefore \quad S_3 = R\sqrt{3}.$$

XVII. *The side of an inscribed equilateral triangle equals the radius of the circle multiplied by* $\sqrt{3}$.

Considering next the inscribed square, let the side be S_4 in Fig. 161. Then in the right triangle AOB each leg AO, BO, equals the radius R and the hypotenuse $AB = S_4$. Therefore

$$S_4{}^2 = R^2 + R^2 = 2R^2$$

$$\therefore \quad S_4 = R\sqrt{2}$$

XVIII. *The side of an inscribed square equals the radius multiplied by $\sqrt{2}$.*

83. Remarks on Inscription of Regular Polygons and the Division of the Circle. In the preceding articles we have inscribed the regular polygons of 3, 4, 5, 6, 8, 10, 12, and 15 sides, and from these others can be obtained by repeated doubling of these. It is not possible, however, to inscribe by simple geometrical methods the regular polygons of 7, 9, 11, 13 and 14 sides (to mention only those in the list of article 79), or those obtained by doubling these.

The regular polygon of 17 sides can be inscribed directly but the construction and proof are too complicated to give here. The method and its proof were discovered by the great German mathematician Gauss in 1796 when he was only 19 years old. When this polygon is inscribed then of course those of 34, 68, etc., sides can be inscribed. Gauss also showed that any n-gon of $n = 2^a + 1$ sides, when a is a whole number and n a prime number, can also be inscribed, with those obtained from these by doubling.

There are several very closely approximate methods used by draftsmen for inscribing regular polygons with the usual drawing instruments. Also, by methods of calculation developed in trigonometry the length of the side of any desired inscribed polygon can be determined as closely as desired and thus drawn as accurately as is usually necessary. Similarly, by laying off central angles with a protractor polygons of any number of sides may be drawn as accurately as the protractor can be read. These methods are not mathematically precise, however.

84. Measurement of the Circle. By the expression "measurement of the circle" is meant the determination of the relations between the radius, diameter, circumference and area of any circle. In this article we develop the propositions or rules setting forth these relations. We shall find that they depend on the relations between the circle and its inscribed regular polygons.

XIX. *The circumferences of two circles have the same ratio as their radii.*

In Fig. 164 let (a), (b) represent any two circles, and let their circumferences be C, C' and their radii R, R'. The theorem then states that $C:C' = R:R'$, and this is to be proved.

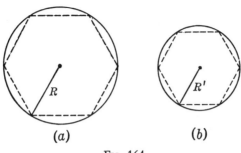

Fig. 164.

Inscribe in the circles two regular polygons of the same number of sides, and let P, P' be their perimeters. Then, according to IV, 80, $P:P' = R:R'$.

Double the number of sides of each inscribed regular polygon, and conceive this process to be repeated, and continued indefinitely, the polygons continuing to be similar. Then P, P' approach C, C' as limits (VII, 80).

But always $P:P' = R:R'$. In the limit, therefore, $C:C' = R:R'$, as was to be proved.

From this result we obtain the fundamental property of the circle. Thus by multiplying the numerator and denominator of the fraction (or ratio) $\frac{R}{R'}$ by 2, which does not change its value, we have $\frac{2R}{2R'}$, or $\frac{D}{D'}$, where D and D' are the diameters of the two circles. The proportion proved in the last proposition becomes therefore $\frac{C}{C'} = \frac{D}{D'}$, or by alternation (III, 61),

$$\frac{C}{D} = \frac{C'}{D'}.$$

Now C/D is the ratio of the circumference to diameter for circle (a), Fig. 164, and C'/D' the same ratio for circle (b). The equation states that this ratio is the *same for both circles*. As these represent *any* two circles, the same is true for any two and therefore for *all* circles, and every circle.

We have, therefore, the extremely important result that

XX. *The ratio of the circumference of a circle to its diameter is constant, i.e., the same for all circles.*

The value of this constant ratio is denoted by the Greek letter π (properly pronounced as "P," but more often in English pronounced as "pie"). That is

$$\frac{C}{D} = \pi; \quad \text{and} \quad \text{hence} \quad C = \pi D, \quad \text{or} \quad C = 2\pi R,$$

since $D = 2R$. In words these relations may be expressed as

XXI. *The circumference of a circle equals the diameter multiplied by π or the radius by 2π.*

Of course this relation is also equivalent to "the diameter is the circumference divided by π."

The value of the important number π will be found in the next article.

XXII. *The area of a circle equals half the product of its radius and circumference.*

Let Fig. 165 represent any circle having the radius $OD = R$; let C be the circumference and A the area. We then have to show that $A = \frac{1}{2}CR$.

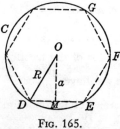

FIG. 165.

Inscribe in the circle any regular polygon. Then $OM = a$ is the apothem, and if P be the perimeter its area is $A' = \frac{1}{2}Pa$ (VIII, 80).

Double the number of sides of the inscribed regular polygon, and conceive this process to be repeated, and continued indefinitely. Then the values of a, P, A' approach as limits R, C, A (VI, VII, 80) but always $A' = \frac{1}{2}Pa$. Ultimately, therefore,

$$A = \frac{1}{2}CR. \qquad \text{—Q.E.D.}$$

This formula can be expressed in terms of R alone by means of the relation $C = 2\pi R$. Thus, substituting this value for C, $A = \frac{1}{2}(2\pi R) \times R = \pi R^2$.

Also $R = \frac{1}{2}D$ and hence $R^2 = \frac{1}{4}D^2$. Therefore, in terms of D, the area formula becomes $A = \frac{1}{4}\pi D^2$. In words,

XXIII. *The area of a circle equals the square of its radius multiplied by π or the square of its diameter by $\frac{1}{4}\pi$.*

From the relations $C = \pi D$ and $C = 2\pi R$, we have also, as indicated following XXI above, $D = C/\pi$ and $R = C/2\pi$.

From the relations $A = \pi R^2$ and $A = \frac{1}{4}\pi D^2$, we have also

$$R = \sqrt{\frac{A}{\pi}} \quad \text{and} \quad D = 2\sqrt{\frac{A}{\pi}}.$$

The relations found in this article are usually stated and explained or illustrated in arithmetic but not strictly proved.

85. Note on the Number π. Although the Greek letter π is used as a symbol to represent the ratio of the circumference of a circle to its diameter, and although the ancient Greeks were the first to investigate thoroughly the relation of this ratio to the properties of the circle and to determine a very closely approximate value of the ratio, they did not use the letter π for the number. It was first used by an English mathematician, William Jones, in 1706.

This choice was made on the following grounds. If the diameter of a circle is unity (1 unit) the circumference is $3.14159 + \ldots$. In the time of the ancient Greeks and until comparatively recent times the word "periphery" was used instead of "circumference," the Greek word being περιφέρεια (*periphereia*), and the initial letter π. Thus if the diameter is 1 the "periphery" is the value represented by π.

Nowadays the word "perimeter" is used for the length of the boundary of a figure formed by *straight* lines (a polygon), "periphery" is used for any *curved* boundary other than the circle (as an oval) and "circumference" for the circle (round).

For the ratio of circumference to diameter the ancient Babylonians used the value 3, as did also the ancient Hebrews (see I Kings, chap. 7, verse 23).

There is an interesting relation between the Anglo-Saxon and Old English words for "yard" and the value of π which may indicate that those people also used the value 3. Thus the Anglo-Saxon words

$$\begin{aligned}
gyrdel &= \text{"girdle" (noun)} \\
gyrdon &= \text{"gird" (verb)} \\
geard &= \text{"yard" (enclosure)} \\
\left.\begin{matrix}gierd\\gyrd\end{matrix}\right\} &= \text{"yard" (measure)}
\end{aligned}$$

Thus a "yard" as an enclosure is a space "girt" by a fence, and if the circumference of a circle were *three* times its diameter, then the girth of a round yard would be three times the distance across it, and for a circle of diameter one *foot* (length unit derived from the human foot) the circumference (girth) would be 3 feet or one "yard" (measure).

The value given for the ratio in the ancient Egyptian manuscript of Ahmes (article 2) is, in modern notation, 3.1604. No calculation survives for this value, however.

The Greek mathematician and scientist Archimedes (article 4) was the first to consider the matter scientifically and leave a record of his work. He proved that the value of the ratio is between $3\frac{10}{71}$ and $3\frac{10}{70}$, or in decimals (to four places) π is between 3.1408 and 3.1428. His method shows, when his calculations are carried further, that to five places $\pi = 3.14159$, or to four places, 3.1416.

The Egyptian Greek astronomer Ptolemy (article 5) seems first to have reduced Archimedes' value to 3.1416; and this value was also calculated by the Hindu Arya-Bhatta (article 6). Other values very nearly correct were given by ancient Hindu and Chinese mathematicians, and a Chinese named Chung-chih gave the value $\frac{355}{113}$ in the fifth century of the Christian era, but did not show how he obtained it. Many other calculations of approximate values were made by the Chinese, Hindus and Arabians in the Middle Ages but none were better than this.

In modern times the first advance in determining π was made by one Adriaen Anthonisz of Holland (lived 1527–1607) who carried out the complete calculation of the Chinese value $\frac{355}{113}$. His son Metius published the calculation in 1625 and this value is now known as the Metius value of π. It is correct to 6 decimal places. Very soon after this time other advances were made by other Europeans. François Vièta (1540–1603) in France found the value of π to 9 decimal places; in Holland Adriaen van Rooman (1561–1615) carried it to 17 places and Ludolph van Ceulen (1540–1610) to 35. These values were all found by geometrical methods based on that of Archimedes.

After the discovery of the branch of higher mathematics called the *calculus* new methods were developed for the calculation of π (see the CALCULUS of this series), and in Germany it was carried to 140 decimal places by George Vega (who died 1793), to 200 places by Zacharias Dase (died 1844), and to 500 by Richter (died 1854). More recently the value of π has been found by William Shanks, in England, to 707 decimal places, which he reached in 1873. In 1961 the value of π was carried out on an IBM 7090 computer to 100,265 decimal places. This operation, which was done by David Shanks and John W. Wrench Jr., required 8 hours 43 minutes.

86. Illustrative Examples.

Example 1. Find the apothem and area of the regular octagon inscribed in a circle of radius one foot.

Solution. According to VIII, 80, the area is half the product of the apothem and perimeter. Since the length of each side was found to be 9.18 the perimeter is $P = 8 \times 9.18 = 73.46$ inches. In order to find the area we must now find the apothem.

Referring to Fig. 159, the radius is $R = \overline{AO} = 1$ ft. $= 12$ in. and in the right $\triangle AMO$ the leg $\overline{AM} = \frac{1}{2}S_8 = \frac{1}{2}(9.18) = 4.59$. Then $\overline{OM}^2 = \overline{AO}^2 - \overline{AM}^2 = (12)^2 - (4.59)^2 = 144 - 21.42 = 122.58$. The apothem is then $\overline{OM} = \sqrt{122.58} = 11.07$ in. $= a$.

Having now the apothem and the perimeter, the area is $A = \frac{1}{2}aP = \frac{1}{2}(11.07 \times 73.46) = 406$ sq. in.

Example 2. A cow is tied to a stake by a rope 15 feet long. Over how much ground can she graze?

Solution. This requires simply the calculation of the area of a circle of 15 feet radius, and as in the last example $A = \pi R^2 = \pi(15)^2 = 3.1416 \times 225 = 706.9$ or about 707 sq. ft.

According to XXIII, 84, this is also $\frac{1}{4}\pi D^2 = .7854(30)^2 = .7854 \times 900 = 706.9$ sq. ft.

Example 3. Suppose it is desired to have the cow in the preceding example graze over 1000 sq. ft. How long must the rope be?

Solution. Here the area of the circle is known and it is required to find the radius, which is the length of the rope. Since $A = \pi R^2$ then $R^2 = \dfrac{A}{\pi}$ and hence $R = \sqrt{\dfrac{A}{\pi}} = \sqrt{\dfrac{1000}{3.1416}} = \sqrt{318.31} = 17.85$ ft. The rope is therefore very nearly 18 feet long.

Example 4. A cow is tethered at the end of a 50-foot rope, which is fastened to a corner of a barn. If the barn is 25×60 feet, over how much area may the cow graze?

Solution. In Fig. 166 let the rectangle $ABCD$ represent the barn, and suppose the rope tied at A. The area over which the cow may graze is then the shaded area.

This area consists of the quadrant *AEF* and the semi-circle *FGH*, having the full rope length $AE = AF = AH = 50$ ft., as radius; and

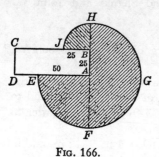

Fig. 166.

of the quadrant *BHJ*. As the rope is stretched around the corner at *B*, the length $BJ = 50 - \overline{AB} = 50 - 25 = 25$ ft. The radius of the small quadrant is therefore $BH = BJ = 25$ ft.

The total shaded area is therefore three quadrants of a circle of 50 ft. radius and one quadrant of 25 ft. radius, and is found by adding $\frac{3}{4}$ of the area of the large circle and $\frac{1}{4}$ the area of the small one. The calculation can be expressed in a single formula, however, as follows.

Let the 50 ft. = R, then 25 ft. = $\frac{R}{2}$. The three quadrants of the large circle are then $\frac{3}{4}(\pi R^2) = 3 \times \frac{\pi R^2}{4}$, and the quadrant of the small circle is $\frac{1}{4}\left[\pi\left(\frac{R}{2}\right)^2\right] = \frac{1}{4} \times \frac{\pi R^2}{4}$. Adding, the total area is

$$A = 3 \times \frac{\pi R^2}{4} + \frac{1}{4} \times \frac{\pi R^2}{4} = \left(3 + \frac{1}{4}\right) \times \frac{\pi R^2}{4} = \frac{13}{4} \times \frac{\pi R^2}{4} = \frac{13}{16} \times \pi R^2.$$

This is the single formula for the total area. The numerical computation is now very simple. Since $R = 50$, $R^2 = 2500$, and the area is $A = \frac{13}{16} \times 2500\pi = \frac{13}{4} \times 625\pi = 13 \times 625 \times \frac{1}{4}\pi = 13 \times 625 \times .7854 = 6382$ sq. ft.

The area could of course have been calculated by finding the two parts of the area separately and adding them. In cases of several calculations of the same kind by formula, however, it usually reduces the calculations required and lessens the likelihood of error if the procedure is first condensed as simply as possible into a single formula, and the computation then performed all at once. This method is very generally used in the applications of mathematics to scientific and technical computations.

Example 5. A semi-circle is drawn on each side of a right triangle as diameter. Show that the area of the semi-circle on the hypotenuse is the sum of those on the legs.

Solution. Figure 167 shows the construction, the semi-circle *ADB* having a diameter $AB = c$, semi-circle BEC having diameter $BC = a$, and semi-circle AFC having diameter $AC = b$.

According to XXIII, 84, therefore, the area $ADB = \frac{1}{2}\left(\frac{\pi}{4}c^2\right)$, area $BEC = \frac{1}{2}\left(\frac{\pi}{4}a^2\right)$, and area $AFC = \frac{1}{2}\left(\frac{\pi}{4}b^2\right)$. The sum of the two semi-circles on the legs is therefore $\frac{1}{8}\pi a^2 + \frac{1}{8}\pi b^2 = \frac{1}{8}\pi(a^2 + b^2)$.

But in the right triangle ABC, by the Pythagorean Theorem, $a^2 + b^2 = c^2$, and hence the sum of the areas of the semi-circles on the legs is $\frac{1}{8}\pi c^2$. And this is the area of the semi-circle on the hypotenuse, as was to be shown.

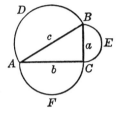

Fig. 167.

This is one of the extensions of the *Pythagorean Theorem* mentioned in connection with the demonstration of XIV, 74, as made by Euclid.

Example 6. In Fig. 168(a) the small semi-circles ACO and BDO are constructed with the radius $AO = BO$ of the large circle as diameter. Show that the shaded area is half the area of the large circle, and that the curved line $ACODB$ is half the circumference.

Solution. Let $AB = d$ be the diameter of the (given) circle. Its area is then $\frac{1}{4}\pi d^2$ and half the area is $\frac{\pi d^2}{8}$. Then $AO = BO = \frac{d}{2}$

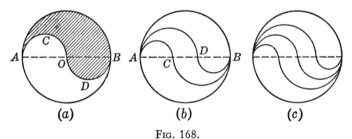

Fig. 168.

is the diameter of each small semi-circle and the area of each is $\frac{1}{2}\left[\frac{1}{4}\pi\left(\frac{d}{2}\right)^2\right] = \frac{\pi d^2}{32}$. The shaded area is then the large semi-circle plus the small shaded semi-circle BDO minus the small semi-circle ACO. But these small semi-circles are equal, so the shaded area is the same as the large semi-circle. That is, it is $\frac{\pi d^2}{8} + \frac{\pi d^2}{32} - \frac{\pi d^2}{32} = \frac{\pi d^2}{8}$.

The circumference of the large circle is $C = \pi d$. Also the semi-circle $ACO = BDO$ has for its diameter half the large diameter, $d/2$. Each small semi-circumference is therefore $\frac{1}{2}\left(\pi \times \frac{d}{2}\right) = \frac{1}{4}\pi d$, and the sum of the two, which is the curve $ACODB$, is $2 \times \frac{1}{4}\pi d = \frac{1}{2}\pi d$, which is $\frac{1}{2}C$. That is, the three curved lines joining A to B are equal.

Similarly it may be shown that if the diameter AB be divided into three equal parts at c, D (Fig. 168(b)), and semi-circles drawn on the segments AC, BC, AD, BD as shown, the area of the circle is trisected and the four curves joining AB are equal.

Likewise the diameter of a circle may be divided into four equal parts, as in (c), and semi-circles drawn on the segments as shown, and the circle is quadrisected and all the curves joining the ends of the diameter are equal. The same is true for five, or any number, of such divisions.

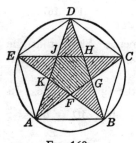

Fig. 169.

Example 7. Inscribe a regular pentagon in a circle (XV, 81) and draw all the diagonals, as in Fig. 169. Show that the sum of the angles at the vertices of the resulting five-pointed star equals two right angles.

Solution. Since each vertex $\angle ADB$, BEC, etc., is inscribed in an arc ($\widehat{AB}$, $\widehat{BC}$, etc.) which is $\frac{1}{5}$ of the circle or $72°$, each of these angles is therefore $36°$ (XVII, 56).

The sum of the five vertex angles is therefore $5 \times 36 = 180° = 2$ rt. angles, the same as the sum of the angles of any triangle.

It will be observed that this five-pointed star is the *pentagram* of the Pythagoreans (article 3). It has many other interesting properties beside the one worked out here.

87. Exercises and Problems.

1. Find by calculation and check by the protractor the angle subtended at the center of the circumscribed circle by each of the regular polygons of article 79.

2. The side of a certain regular polygon is 6 inches and the side of another of the same number of sides is 2 inches. How are their areas related?

ART. 87   REGULAR POLYGONS AND THE CIRCLE

3. The side of a regular octagon is 10 inches and the apothem is 12.07 inches. What is the area?

4. A regular hexagon is inscribed in a circle of one foot diameter. What is the area of the hexagon?

5. What is the difference between the areas of a regular pentagon and a regular decagon if the side of each is 10 inches?

6. What is the difference between the areas of a regular undecagon and dodecagon if the side of each is 10 inches?

7. The diameter of a circle is 20 inches. What is the length of the side of the inscribed regular pentagon; decagon; pentadecagon?

8. What are the sides of the regular polygons of 20 and 30 sides inscribed in the circle of the preceding problem?

9. Find the circumference of a one-foot circle.

10. Find the circumference of a four-inch circle.

11. How wide must a piece of sheet tin be cut to roll into a length of six-inch stove pipe if about a tenth of an inch is allowed for the seam?

12. A regular hexagon is inscribed in a circle of 8 inches radius. What is the area included between the circle and the perimeter of the hexagon?

13. If in Fig. 158 the radius of the circle is one inch, find the

  (1) circumference and area of the circle;
  (2) apothem and side of the inscribed square;
  (3) apothem and side of the inscribed octagon;
  (4) areas of square and octagon;
  (5) area between octagon and circle;
  (6) area between octagon and square.

14. What is the ratio of the area of any circle to that of the inscribed square?

15. What is the ratio of the area of any circle to that of the circumscribed square?

16. A tank may be filled by water flowing at the same speed through a one-inch pipe or a two-inch pipe. How much more time than the larger pipe would the smaller pipe require?

17. The diameter of a grapefruit is four inches and the rind is one-quarter inch thick. A piece is sliced off just grazing the inner flesh. Find the diameter and circumference of the piece sliced off.

18. How many times does a 28-inch wheel turn over in rolling a mile?

19. If the limit of safety for the rim speed of a grinding stone is 5500 feet per minute, what is the diameter of the largest wheel which can safely run at 1500 revolutions per minute?

20. What is the area of a four-foot walk around a circular pond 20 yards across?

21. What is the total force on the piston of a 15-inch steam-engine cylinder when the steam gauge indicates 120 lb. per sq. in.?

22. If two coins, such as half dollars, were melted and made into a single coin of the same thickness, how would the diameter compare with that of the original coins?

23. Find the "angular speed" in degrees per second of a 30-inch car wheel when the car is going 40 miles per hour.

24. Taking the diameter of the earth as about 8000 miles, find the speed in miles per hour (due to the rotation) of a point on the equator, and of a point at latitude 30 degrees.

25. A blade of an airplane propeller is 5 ft. long and it rotates at 3600 revolutions per minute. At what speed in feet per second, and also in miles per minute, does the tip of the blade travel around its shaft?

Chapter 9

SURFACE AREAS AND VOLUMES OF SOLIDS HAVING PLANE SURFACES

88. Lines and Planes in Space. So far we have discussed only those geometrical figures which lie in or are drawn upon a plane. The geometry of such figures is called *plane* geometry or geometry of *two dimensions* (article 15).

The elements or fundamental entities with which we deal in plane geometry are *points* and *lines*, and from these are formed angles, polygons, circles, etc. These figures do not occupy or fill up space.

We now go one step further and consider geometrical figures which do occupy space. Such figures are called *solids* and their geometry is called *solid* geometry or geometry of *three dimensions* (article 15).

The elements or fundamental entities with which we deal in the treatment of solids are points, lines and *surfaces* and from these three are formed all geometrical solids. As we saw in article 15 surfaces are of two kinds, *plane* and *curved*. In this chapter we shall consider only plane surfaces.

In beginning the study of this chapter the reader should now turn to article 15 and read it again.

In a plane or in space any two distinct points determine a straight line, and two straight lines can intersect in only one point (article 22). Consider now a straight line in space. A plane may pass through this line and can turn about it as an axis, as illustrated by a piece of cardboard (which actually has thickness, however) held at two fixed points on its edges and turned about these points. If, however, a third point be fixed in the plane outside the line joining the first two fixed points, the plane cannot be turned about the line. Thus the plane is fixed in position by the line and the point outside the line.

But *any* two distinct points in the original line determine it. These two and the third point, not in the line, therefore, also fix the plane.

Similarly one plane can contain two intersecting or two parallel lines. For the plane could turn about any one line, but if it must also pass through the other fixed line, either parallel to or intersecting the first, then it cannot turn but must remain in one fixed position.

The last two paragraphs may be summarized as follows:

I. *A plane in space is determined by*
   (1) *a straight line and a point outside it;*
   (2) *three points not in the same straight line;*
   (3) *two intersecting straight lines; or*
   (4) *two parallel straight lines.*

Condition (2) is illustrated by the fact that a chair or table with three legs will sit steadily on any floor but in the case of four legs three will sit steadily while the fourth may or may not, as its foot may be outside the plane of the first three.

Since a curved line cannot lie wholly in a plane in every position into which it may be turned but a straight line can (article 15), then a plane can turn about a straight line into any position, and

II. *The intersection of two planes is a straight line.*

If a straight line pierces a plane at a point and is perpendicular to every line in the plane which passes through that point, the line and the plane are said to be perpendicular. But since there can be only one perpendicular from a point outside the plane to each of these lines in the plane (X, 29) then

III. (1) *At a point in a plane or from a point outside the plane there can be only one line perpendicular to the plane;*

and therefore, also,

III. (2) *Through a point outside a line there can pass only one plane perpendicular to the line.*

In dealing with lines in a plane an angle, or more specifically, a *plane angle* is taken as the opening or difference of direction between two intersecting lines. The lines are called the *sides* of the angle, and their point of intersection its *vertex*. The *measure* of the angle is the difference in direction or the inclination of its sides.

Similarly two intersecting planes may have different inclinations, and we define:

A *dihedral angle* is the opening between two intersecting planes.

ART. 88 SURFACE AREAS AND VOLUMES OF SOLIDS   205

The two planes are called the *faces* of the dihedral angle, and their line of intersection is its *edge*.

The *measure* of a dihedral angle, in degrees, say, is the same as the plane angle formed by two lines, one in each face, which are perpendicular to the edge at the same point. This is called the *plane angle of the dihedral angle*.

Thus imagine a piece of paper folded along a straight line and the two parts allowed partly to separate rather than lying flat together; they then form a dihedral angle. If a line be drawn on each part of the paper so that both pass through the same point on the edge of the dihedral angle (the line of the crease) and both are perpendicular to the edge, then the ordinary plane angle formed by these two lines is the measure of the dihedral angle, and is called its plane angle.

Figure 170 represents a dihedral angle formed by the two inclined planes $ABCD$ and $ABEF$. These planes are the faces, and the line AB is the edge of the dihedral angle. If the lines BC, BE are thought of as both perpendicular to AB then the ordinary (plane) $\angle CBE$ is the measure of the dihedral angle.

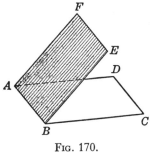

FIG. 170.

A dihedral angle is designated by naming its edge or the edge and the face planes. Thus in Fig. 170, dihedral $\angle AB$ or dihedral $\angle(AC\text{–}AB\text{–}AE)$. (The dashes are not minus signs but simply separate the pairs of letters.)

If three intersecting planes all pass through one point, as three sides of a box meeting at a corner, they are said to form a *trihedral* angle. The three planes are the *faces* of the trihedral angle; the three lines in which they intersect in pairs are its *edges;* the point through which they all pass is its *vertex;* and the three plane angles formed by the intersecting edges at the vertex are the *face angles* of the trihedral angle.

Figure 171 represents a trihedral angle. The triangular figures AVB, BVC, CVA represent portions of the three planes which meet in the point V. These planes are the three faces of the trihedral angle and V is its vertex. The lines AV, BV, CV, in which the planes intersect in pairs, are the edges of the trihedral angle. The three plane angles $\angle AVB$, $\angle BVC$, $\angle CVA$ are the face angles of the trihedral angle at V

Three *or more* planes which in the same way pass through one point and intersect two by two in straight lines are said to form a *polyhedral angle*. The polyhedral angle of any number of faces and edges is described in the same manner as the polyhedral angle of three faces (trihedral). A polyhedral angle is also sometimes called a *solid angle* in contradistinction from a *plane* angle.

Dihedral, trihedral, polyhedral angles, etc., are named from the Greek and Latin roots *di* ("two"), *tri* ("three"), *poly* ("several"), and *edra* ("base" or "face").

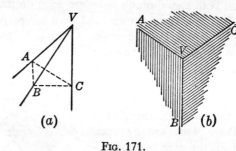

Fig. 171.

If the plane angle of a dihedral angle is a right angle the faces are perpendicular and the dihedral angle is called a *right dihedral angle*. The two sides of a square or rectangular box meeting at an edge form a right dihedral angle.

If the face angles of a trihedral angle are plane right angles the trihedral angle is said to be a *right trihedral angle*. Thus the three sides of a square or rectangular box meeting at a corner form a right trihedral angle (see Fig. 171(*b*)).

If the edges *AV*, *BV*, *CV* in Fig. 171 spread farther apart (as the legs of a tripod) the face angles at *V* become larger and larger and the three plane faces of the trihedral angle "flatten out," so to speak. If this opening or flattening continues the three planes will finally lie flat together and coincide as one plane, and the figure is then no longer a trihedral angle. The three edges are then simply lines radiating from a point in a plane, and the sum of the three original face angles *AVB*, *BVC*, *CVA* is simply the sum of the plane angles about a point, which is four right angles or 360 degrees (VI, 26).

From this we have the result that *the sum of the face angles must be less than* 360 *degrees* if the figure is to be a trihedral angle. Similarly a polyhedral angle of any number of faces may flatten out into a plane, when the polyhedral angle ceases to be an angle, and the sum of its face angles becomes the sum of the plane angles about a point in a plane. Therefore

IV. *The sum of the face angles of any polyhedral angle must be less than 360 degrees.*

This means that the sum can never equal 360°, for then the figure would not be formed of several intersecting planes but of only one.

Lines and planes in space and dihedral and polyhedral angles formed by them possess many interesting properties analogous to those of points, lines and plane angles studied in Part II, but we shall not undertake a detailed study of them. We now pass to the consideration of figures formed of lines, planes, dihedral and polyhedral angles in space.

89. Solids. A *solid figure* has already been defined (articles 15, 88) in terms of the fundamental elements of geometry (points, lines, planes, etc.) but in order to be able to visualize such a figure more clearly we may repeat that a solid figure is one which encloses a portion of space, as illustrated by a box, a block of stone, a tin can, a globe or ball, and the like. Such a figure may be "solid" or "hollow" in the ordinary sense of these words, and it may be closed or open, filled or empty. In all cases it will be called a *solid* to distinguish it from a *plane* figure, and it will be studied as if it were as "solid" as a block of stone.

There are many different forms of solids. They may be classified as those which have plane surfaces and straight edges, and those which do not. The surfaces of the first kind of solid may be any of the plane figures studied in Part II. The surfaces of the second kind of solid may be curved in any way whatever and if it has any "edges" these may be any form whatever of curved line. Altogether the number of forms of solids is unlimited.

Of this great number of possible solids, however, only a few are of common interest and practical importance, and only these will be studied here.

In this chapter we will take up the study of the first form of solid mentioned above, those having plane surfaces and straight edges. These will not be studied in all their details, however. Thus in Part II the circle was studied in Chapters 5 and 8. In Chapter 5 the *general* properties of the circle were discussed while in Chapter 8 only its numerical properties were considered, or, as these are sometimes called collectively, the *mensuration* or measurement of the circle. In this and the following chapters we shall likewise consider only those properties of solid figures which are involved in their measurement.

90. Solids with Plane Surfaces. Since planes intersect only in straight lines (II, 90) solids having plane surfaces can only have straight edges. On this account solids having plane surfaces are called *rectilinear* solids.

The different portions of a surface of a rectilinear solid, each of which is a single plane figure, are called the *faces* of the solid. A rectilinear solid is named according to the shape or number of faces it may have. Of all the rectilinear solids those whose faces are rectangles are the most important, and these are called *rectangular* solids. Of all the rectangular solids, that one whose faces are equal squares is the most important; it is called the *cube*.

If the faces of a rectilinear solid are regular polygons, all of which are congruent, it is called a *regular* solid.

Other forms of rectilinear solids are described in later articles.

As in the case of plane figures, the lengths of the edges of a rectilinear solid, and of other distinguishing or characteristic lines on the surface or inside the figure, are called its *dimensions*. The area of its faces, or the sum of those areas, is called its surface area, and the amount of space which it occupies or encloses is called its *volume*. Volume and its measurement are discussed in more detail in the next article.

The characteristic features of solids and the relations between their dimensions, areas and volumes are collectively referred to as their *properties*. The properties having most to do with the *measure* of the dimensions, areas and volumes are called the *numerical properties* of a figure. In this chapter we take up the numerical properties of a few rectilinear solids.

91. Rectangular Solids and Volume. A *rectangular solid* is a solid which has six plane faces, each of which is a rectangle. Figure 172 represents a rectangular solid. Only three of the six rectangular faces of a rectangular solid can be seen at any one time. In Fig. 172 these three visible faces are the rectangles $ABCD$, $BCFG$ and $CDEF$. These three faces may be called the end, side and top, respectively.

A rectangular solid has three dimensions; its *length*, *width* (or breadth) and *altitude* (depth, height or thickness). In Fig. 172 $BG = CF = DE$

ART. 91 SURFACE AREAS AND VOLUMES OF SOLIDS

is the length, and may be represented by l; $AB = DC = EF$ is the width, represented by w; and $AD = BC = GF$ is the altitude h. The lines named are also called the *edges*, of which the figure has twelve, three being invisible in any view. The rectangular face on which the solid is supposed to stand is called its *base*.

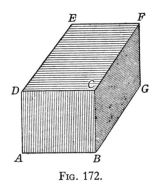

FIG. 172.

Of the six faces there are three equal pairs: top and bottom, two sides, and two ends. According to IV, 72, the area of the top is the product of length and width (lw), and of the top and bottom, twice this product. The area of either side is the product of length and altitude (lh), and of both sides twice this. Similarly the area of the two ends is twice the product of width and altitude (wh). The total surface area is the sum of these double products, or double the sum of the three single products. That is,

V. *The total surface of a rectangular solid is twice the sum of the three products: length by width, width by altitude, altitude by length.*

In symbols this is

$$A = 2(lw + wh + hl).$$

Since the opposite faces of a rectangular solid are parallel planes and its edges form three groups of four parallel lines, a rectangular solid is also called a *parallelepiped*.

The *volume* of a solid has been defined as the amount of space which it occupies or encloses. The *unit of volume* is a rectangular solid each of whose three dimensions is the unit of length. Such a solid is called a *unit cube*, and its volume is called a *cubic unit* of volume, such as the cubic inch, centimeter, foot, etc. The numerical *measure of the volume* of a solid, called simply *the volume* of the solid, is the number of unit cubes which it contains.

The method of calculating the volume of a rectangular solid from its dimensions is easily described and understood, and is usually described and illustrated in arithmetic. We will here develop something of the geometrical properties on which the rule of calculation is based. We begin by showing that

VI. *The ratio of the volumes of two rectangular solids with equal bases equals the ratio of their altitudes.*

In Fig. 173 let (a) and (b) represent two rectangular solids having equal rectangles as bases, and having volumes V, V' and altitudes $AB = c$, $A'B' = c'$. We then have to show that $V:V' = c:c'$.

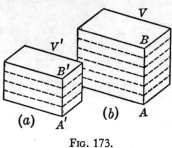

Fig. 173.

If the altitudes AB, $A'B'$ are commensurable, apply their common measure and suppose that it is contained in AB m and in $A'B'$ n times. Then (article 55) the ratio of the altitudes is $c:c' = m:n$.

At the several points of division of the lines AB, $A'B'$ pass planes through the solids perpendicular to the altitudes, parallel to the bases and one another, dividing the solids into "slices" or "slabs" which are all equal, having equal bases and altitudes, the common altitude being the common measure of AB and $A'B'$.

There are then m of these smaller solids of equal volume v in the solid of volume V; hence $V = mv$. There are n of them in V'; hence $V' = nv$. Therefore

$$\frac{V}{V'} = \frac{mv}{nv} = \frac{m}{n}.$$

But we found also that $\frac{c}{c'} = \frac{m}{n}$. Therefore $\frac{V}{V'} = \frac{c}{c'}$ as was to be proved.

If the altitudes are incommensurable the ratio is found to be the same, by the method of limits, as in the proof of XV, 56, and the theorem is completely proved.

VII. *Two rectangular solids of equal altitudes have the same ratio as their base areas.*

In Fig. 174 let (a), (b) be the two solids, having volumes V, V'; bases of lengths a, a' and widths b, b'; and equal altitudes c. The theorem then states that $\frac{V}{V'} = \frac{ab}{a'b'}$, the products ab and $a'b'$ being the base areas.

Art. 91 SURFACE AREAS AND VOLUMES OF SOLIDS

Construct as in (c), a third rectangular solid having the same altitude c as the first two; and having the same length a as V and the same width b' as V'; and having volume V''.

Then since V'' has the two dimensions a, c the same as V, one face having these as edges can be taken as base and b, b' will be the altitudes. According to VI above, therefore, $\dfrac{V}{V''} = \dfrac{b}{b'}$.

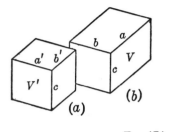

Fig. 174.

Also, V'', V' have the same two dimensions b', c; then in the same way we have $\dfrac{V''}{V'} = \dfrac{a}{a'}$.

Multiplying these two equations together we obtain

$$\frac{V \times V''}{V'' \times V'} = \frac{a \times b}{a' \times b'},$$

or, cancelling the V'' in numerator and denominator, $\dfrac{V}{V'} = \dfrac{ab}{a'b'}$, as was to be proved.

VIII. *Any two rectangular solids are to each other as the products of their three dimensions.*

Let one of the two rectangular solids have the three dimensions a, b, c and volume V, and the other dimensions a', b', c' and volume V'. Construct also a third rectangular solid having dimensions a, b', c and volume V''.

Then since V'' has one dimension (b') the same as one of V' (take this as the altitude) we have as in VII above $\dfrac{V''}{V'} = \dfrac{ac}{a'c'}$.

Also, V'' has two dimensions (a, c) the same as V (let these form the equal bases); hence, as in VI, $\dfrac{V}{V''} = \dfrac{b}{b'}$.

Multiply these two equations together, cancelling the V'', and

$$\frac{V}{V'} = \frac{abc}{a'b'c'}. \qquad \text{—Q.E.D.}$$

By means of the last proposition we now prove the following fundamental and extremely important proposition:

IX. *The volume of any rectangular solid equals the product of its three dimensions.*

Let the dimensions of the rectangular solid be a, b, c and its volume V. Construct also a *unit cube* whose dimensions are $a' = 1$, $b' = 1$, $c' = 1$ length unit, and whose volume is, therefore, by definition, $V' = 1$ cubic unit.

Then as in VIII, $V/V' = abc/a'b'c'$. Putting in the values of a', b', c', V' this becomes $\frac{V}{1} = \frac{a \times b \times c}{1 \times 1 \times 1}$, or, since $1 \times 1 \times 1 = 1$,

$$V = a \times b \times c = abc. \qquad \text{—Q.E.D.}$$

All volume measurements are based on this proposition. It is simply illustrated as follows.

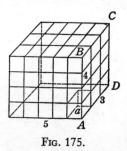

Fig. 175.

Suppose that the dimensions of a rectangular solid are 3, 4 and 5 units, say inches, as in Fig. 175 below, and imagine the solid built of square blocks 1 inch on each edge such as a, shown at the corner A. This little block is 1 *cubic inch*. The section $ABCD$ is then a layer consisting of $3 \times 4 = 12$ of these blocks, or cubic inches, and the whole space occupied by the solid, its *volume*, is filled up by 5 such layers as $ABCD$. There are, therefore, $5 \times 12 = 60$ of these blocks, that is 60 cu. in., in the entire solid. But this $60 = 5 \times 12 = 5 \times 4 \times 3$ is the product of the three dimensions 3, 4 and 5. That is, in general, the volume of the rectangular solid is the product of its three dimensions, as strictly proven above. The description just given as an illustration is the explanation of the volume formula or rule generally given in arithmetic.

If we use the symbols l, w, h for length, width and altitude then the volume of a rectangular solid is $V = lwh$. But lw is the area of the

ART. 92   SURFACE AREAS AND VOLUMES OF SOLIDS   213

bottom or *base*, and since the volume formula can be written as $V = (lw) \times h$, we can state the rule as follows:

X. *The volume of a rectangular solid is the product of the base area and the altitude.*

The first form of the volume rule is more convenient for numerical calculations; the second will be useful in the considerations of the next article.

If all the dimensions of a rectangular solid are equal, its six rectangular faces are all equal squares and the solid is, as we have seen, called a *cube*. The rules for surface area and volume then become

XI. *The surface area of a cube is six times the square of its edge.*

XII. *The volume of a cube is the cube of its edge.*

It is because of the last rule that volume measure is called cubic measure and a cube which has an edge of one *inch* (centimeter, foot, etc.) is called a *cubic inch* (centimeter, foot, etc.).

It is also on this account that the product obtained by using the same number three times as a factor (the third *power* of the number) is called in algebra the *cube* of the number.

92. Prisms and Pyramids. A *prism* is a rectilinear solid two of whose faces are polygons and the rest parallelograms or trapezoids. If the two polygonal faces are congruent and parallel and the remaining faces are rectangles (as many as there are sides to each of the polygons) the prism is called a *right prism*. Fig. 176 represents a right prism.

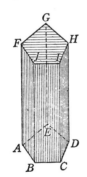

Fig. 176.

In Fig. 176 the parallel polygons are the pentagons *ABCDE* and *FGHIJ*, and the five remaining faces are the rectangles *ABJF*, *BCIJ*, *CDHI*, *DEGH* and *EAFG*. The polygons are called the ends or *bases* of the prism and the rectangular faces its *sides*. A prism may have as few as three sides and any number more. The polygons may have any number of sides and may be regular polygons or not, but the two must be congruent. Thus in Fig. 177 (*a*) represents a triangular prism, having congruent triangles *ABC* and *DEF* as bases and three rectangular faces *ABED*, *BCFE*, *ACFD*.

As seen in Fig. 177(*b*) a right triangular prism may be thought of as formed by splitting a rectangular solid into halves, and the volume of the rectangular solid is the product of its base area and altitude (X, 91). But the prism is half the rectangular solid, its triangular base is half the rectangular base of the rectangular solid, and its altitude is the same. Therefore

XIII. *The volume of a right triangular prism is the product of its base area and altitude.*

Now as seen in Fig. 177(c) any right prism can be divided into triangular prisms, the sum of whose volumes is, therefore, the volume

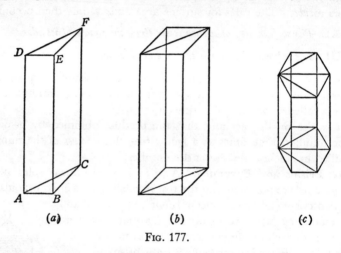

Fig. 177.

of the complete prism, and the sum of whose bases is the base of the complete prism. Therefore, also,

XIV. *The volume of any right prism is the product of its base area and altitude.*

A *pyramid* is a solid which has a polygon as base, and for faces, triangles whose vertices meet in a single point, forming a polyhedral angle. If the polygonal face is regular and the triangles are all isosceles and congruent, the pyramid is a *regular right pyramid*. Fig. 178 represents a regular right pyramid with a pentagonal base, *BCIEF*. The point where the vertices of the triangular faces meet (the vertex of the polyhedral angle) is called the *vertex* of the pyramid, or the *apex*, and the length of the vertical line *AH* from the vertex to the center

ART. 92   SURFACE AREAS AND VOLUMES OF SOLIDS   215

of the base is the *altitude* of the pyramid. The altitude of any one of the triangles forming the faces of a right pyramid, as *AG*, is called the *slant height* of the pyramid. The perimeter of the polygon is also the *perimeter* of the base of the pyramid.

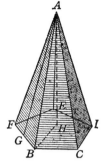

Fig. 178.

The area of any one of the triangular faces is half the product of its altitude and base (VII, 74) and the total area of all the faces is the sum of these half products. Since the altitude is the same for all the triangular faces, the total area is half the product of the triangular altitude by the sum of the triangular bases. But the common altitude is the slant height of the pyramid, the sum of the bases of the faces is the perimeter of the base of the pyramid, and the sum of the face areas is the total face or *lateral* area of the pyramid. Therefore,

XV. *The lateral area of a right pyramid is half the product of its slant height and base perimeter.*

In order to find the volume of a pyramid, consider Fig. 179. Here *F* is the center point of the interior of the cube *AB* with the square base *BCDE*. By drawing the lines *FB*, *FC*, *FD*, *FE*, from *F* to the corners of the base, a regular right pyramid is formed, having the square base *BCDE*, the vertex *F*, and an altitude equal to half the altitude of the cube. By drawing lines from *F* to each of the other four corners of the cube, five other such pyramids will be formed, having the other five faces of the cube as bases and the point *F* as common vertex. Thus the cube will be divided into *six* equal pyramids with the base of each equal to a face of the cube and the altitude of each equal to half that of the cube.

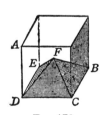

Fig. 179.

Therefore the volume of each pyramid is equal to one sixth the volume of the cube. But the volume of the cube, a rectangular solid, is equal to the product of its altitude by the area of one of its faces (base). Therefore the volume of each pyramid is one sixth the product of its base area (cube face) by the cube's altitude, which is *twice* the altitude of the pyramid. Thus the volume of the pyramid is one sixth the product of its base area by twice its altitude, or

XVI. *The volume of the right pyramid having a square base is one third the product of its base area and altitude.*

If now a plane be passed through the two edges *BF*, *DF* it will cut the base of the square pyramid along the diagonal *BD* and the pyramid will be divided into two equal pyramids having the same altitude and equal triangular bases, each of which is half the square base (XXIX, 46). Therefore, also,

XVII. *The volume of a pyramid having a triangular base is one third the product of its base area and altitude.*

Now by passing planes through any two edges of *any* pyramid having more than three faces (Fig. 178) the planes intersect the polygonal base along the diagonals as *BE*, *CE*, Fig. 178, and so divide the pyramid into other pyramids having triangular bases and the same altitude. The sum of the volumes of these triangular pyramids then equals the volume of the original pyramid, and the sum of their base areas equals the base area of the original pyramid. Therefore,

XVIII. *The volume of any pyramid is one third the product of its base area and altitude.*

Thus the volume in cubic inches of the pyramid of Fig. 178 is found by multiplying the area of the pentagon *BCIEF* in square inches, by the altitude *AH* in inches, and dividing by three.

93. Regular Solids. A general name which includes all rectilinear solids (solids having plane faces and polyhedral angles, or corners) is *polyhedron*. Polyhedrons are named according to their numbers of faces, after the manner of naming polygons from the number of their sides. Thus polyhedrons of 4, 5, 6 sides, etc., are called *tetrahedron*, *pentahedron*, *hexahedron*, etc., respectively.

A pyramid may have four or more faces (including base); a prism may have five or more (including the two bases); and a rectangular solid in particular has six. Thus a pyramid of three faces beside the base is a tetrahedron, and a rectangular solid is a hexahedron.

If all the faces of a polyhedron are congruent regular polygons the solid is called a *regular solid* or *regular polyhedron*. A pyramid having four equal equilateral triangles as base and faces is a *regular tetrahedron*, and a cube, having six equal squares as faces, is a *regular hexahedron*.

We now prove a most remarkable proposition concerning regular polyhedrons, namely:

ART. 93   SURFACE AREAS AND VOLUMES OF SOLIDS   217

XIX. *Only five regular polyhedrons are possible.*

To begin with, a polyhedral angle must have at least three faces (article 88) and the sum of its face angles must be less than 360 degrees (IV, 88).

(1) The equilateral triangle face is the regular polygon of the least possible number of sides, and since each angle of an equilateral triangle is 60° polyhedral angles may be formed by 3, 4 or 5 equilateral triangles, but not by 6, since the sum would then be 360.

Hence *three* different regular polyhedrons are possible with equilateral triangles as faces.

(2) Since each angle of a square is 90° a polyhedral angle may be formed by combining 3 squares, but not 4, since $4 \times 90 = 360°$.

Hence only *one* regular polyhedron is possible with squares as faces.

(3) Since each angle of a regular pentagon is 108° (Example 7, article 86) a polyhedral angle may be formed of 3 regular pentagons but not 4, since $4 \times 108 = 432°$.

Hence *one* regular polyhedron is possible with regular pentagons as faces.

(4) Since each angle of a regular hexagon is $(1 - \frac{2}{6}) \times 180 = 120°$ (XLII, 50), the sum of three such angles is 360° and hence no polyhedral angle can be formed by combining regular hexagons. Similarly no polyhedral angles can be formed with regular heptagons, octagons, etc., as faces.

Hence no other regular polyhedrons are possible than those stated under (1), (2), (3) above.

(5) Therefore only *five* regular polyhedrons are possible, as was to be proved.

The five possible regular polyhedrons (regular solids) are the following:

Faces	Names
(1) Four equil. triangles (triangular pyramid)	Tetrahedron
(2) Six squares (cube)	Hexahedron
(3) Eight equilateral triangles	Octahedron
(4) Twelve regular pentagons	Dodecahedron
(5) Twenty equilateral triangles	Icosahedron

The five regular solids are shown in Fig. 180.

In books on solid geometry and on geometrical and mechanical drawing the five regular solids are constructed as problems in ordinary

Fig. 180.

geometrical construction, such as we have solved in the construction of plane figures.

The five regular solids may be formed of cardboard or stiff paper by drawing and cutting out the figures shown in solid lines in Fig. 181, folding these on the dotted lines, and sticking the free edges together.

The method of folding and fitting is illustrated in Fig. 182, which shows the formation of the icosahedron.

The following are two other remarkable properties of regular (as well as any other) polyhedrons:

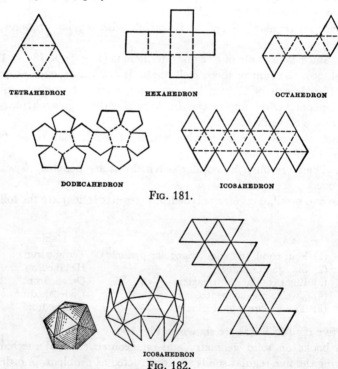

Fig. 181.

Fig. 182.

XX. *In any polyhedron the number of faces plus the number of vertices equals the number of edges plus two.*

XXI. *If a polyhedron has n polyhedral angles (vertices) the sum of all the face angles is $(n - 2) \times 360$ degrees.*

The proofs of these are somewhat involved and will not be given here. They may be found in the textbooks mentioned at the ends of articles 7 and 9.

The regular polyhedrons (solids) are analogous to the regular polygons (plane) in that they may be inscribed in a sphere, as the polygons may be inscribed in a circle.

When the length and the number of the sides of a regular polygon are known its area can be calculated (VIII, 80, or article 76). Since the regular polyhedrons have regular polygons as faces, their surface areas may also be calculated from their dimensions.

Any regular polyhedron may be thought of as made up of similar and equal right regular pyramids (article 92) so placed that their apices (or apexes) are all together at the center of the polyhedron and their bases form the faces of the polyhedron. Since the method of calculating the volumes of the pyramids is known (XVIII, 92), the volume of the regular solid is known. The regular tetrahedron is itself, of course, a regular triangular pyramid, and the cube is shown divided into pyramids in Fig. 179.

The derivations or proofs of the area and volume calculations are somewhat complicated and will not be given here, but the results, the formulas themselves, are given in the following table. In these formulas l represents the length of an edge of the regular polyhedron, which is a side of its regular polygonal face.

Regular Polyhedrons		
Name	Surface Area	Volume
Tetrahedron......	$1.732l^2$	$0.118l^3$
Hexahedron......	$6.000l^2$	$1.000l^3$
Octahedron......	$3.464l^2$	$0.471l^3$
Dodecahedron....	$20.646l^2$	$7.663l^3$
Icosahedron......	$8.660l^2$	$2.182l^3$

94. Note on the Regular Solids. In the development of geometry and measurement the regular polyhedrons have attracted as much attention as the regular polygons. Historians believe that some of them were known to the ancient Egyptian priests and builders. The first geometers to make a systematic study of them and to leave a record of their work, however, were the Pythagoreans (article 3). It had long been thought that there were only three of the regular solids: the tetrahedron, cube and octahedron, formed by means of the equilateral triangle and the square. The Pythagoreans soon discovered the dodecahedron, however, having regular pentagons as faces. They also later discovered the icosahedron.

The Pythagoreans considered the regular solids from the viewpoint of their inscription in a sphere, in analogy with the regular polygons. One of their members, Hippasus by name, is said to have been drowned by the society because he made public the discovery of the "sphere with twelve pentagons" (dodecahedron), the *pentagram* having been their fraternal insigne (article 3 and Example 7, article 86).

Plato and his school (article 3) also studied the regular solids but, unlike the Pythagoreans, did not try to keep their work secret. On account of the fame of Plato the regular solids have been called the "Platonic Bodies." It is not definitely known whether Plato was aware that only five are possible but by the time of Euclid this was known, and Euclid's *Elements* contains a proof that only five regular solids are possible. Whether this proof was original with Euclid is not known but it is the first such proof of which there is definite record. Our proof (XIX, 93) is not Euclid's proof, which was based on inscription in a sphere, but seems first to have been given by the noted English mathematician, Augustus De Morgan (1806–1871).

Euclid gave a systematic and fairly complete treatment of the regular solids, including their geometrical construction, or inscription in a sphere. An Arabian, Abul Wefa (940–998), was the first to show, in medieval times, that each one can be constructed by means of a single opening or setting of the compass (dividers).

The proposition XX, 93, is known as *Euler's Theorem* because the great and famous Swiss mathematician Leonhard Euler (1707–1783) seems to have been the first to give a correct proof of it. It was known, however, to Descartes (article 8) but he left no proof of the theorem.

In recent times the German mathematician, Felix Klein (1849–1925), by the study of the theory of the regular solids, and the icosahedron in particular, has opened up a new method and almost a new branch of mathematics bearing on the theory of certain algebraic equations.

ART. 95   SURFACE AREAS AND VOLUMES OF SOLIDS   221

From the earliest times in the study of the regular solids there has been conjecture among the mystically inclined concerning the possibility of a relation between the properties of these figures and phenomena in nature. It is known that many natural crystals have the forms of various ones of the regular polyhedrons, or certain modified related forms, and other such relations have been diligently sought. Before the discovery of the icosahedron the Pythagoreans thought that the tetrahedron and the octahedron (4 and 8 faces) represented the two so-called "elements" of nature, fire and air, and that the hexahedron and dodecahedron (6 and 12 faces) represented the other two "elements," earth and water. As it was supposed that there were only four such elements they were at a loss as to the significance of the icosahedron when it was discovered, so they assigned it to the universe of nature as a whole. Their position was somewhat like that of the astrologers who used to say that there were and could be only seven planets because the number seven is one of the mystic or perfect numbers. As other planets continue to be discovered, however, they easily extricate themselves from their difficulties. (In this connection see the ALGEBRA of this series.)

In more recent times the famous German astronomer, Johannes Kepler (1571–1630), thought at one time that he had found a relation between the regular solids and the number and distances of the planets, and his supposed discovery attracted much attention, but he later became convinced that there is no such relation.

Although of no great utilitarian importance, except as used in certain art forms, the regular polyhedrons are of the greatest scientific interest and importance, and it is for this reason that we have considered them here.

95. Exercises and Problems.

1. Find the surface area and volume of a rectangular solid which is 8 inches wide, 12 inches long and 6 inches deep.

2. How many square yards of sheet metal are required to make an open rectangular tank 3 feet wide, 4 feet deep, and 8 feet long; and how many gallons will it hold? (One gallon contains 231 cubic inches.)

3. The steel rods in a fence are hexagonal in shape and each face is half an inch wide. The rods are 8 feet long and set 4 inches apart, and the fence extends 40 yards. How many pounds of steel (0.278 lb. per cu. in.) does the fence contain; and in painting the fence how much surface is covered? (Disregard ends of rods.)

4. A church steeple has the form of a tall pyramid. The base is a regular

hexagon 4 feet on a side, and the slant height of the steeple is 30 feet. How many shingles will be required to cover the steeple if the effective area of each shingle is 4 × 6 inches, making no allowance for cutting or fitting?

5. Find the base area, the altitude, and the volume of the pyramidal steeple of the preceding problem.

6. A certain square (cubical) box can hold 12 bushels of grain. If the edge of another square box is twice as long, how many bushels will it hold?

7. If the edge of a square tank is 12 feet, and a second tank is to have one-third its capacity, what must be the edge of the smaller tank; and how is its total surface related to that of the larger?

8. Find the surface and volume of each of the regular solids if the edge of each is 2 inches.

9. The regular tetrahedron, octahedron and icosahedron are formed of equilateral triangles. If the triangles are all equal, what are the ratios of the areas and volumes of the three solids?

10. A cube and a regular tetrahedron have the same volume. What is the ratio of their edges?

11. A regular icosahedron is to be formed from a square foot of paper of suitable shape (as in Fig. 182). What will be the length of the side of each of the triangles?

12. What is the edge of a regular tetrahedron of one square foot surface?

Chapter 10

THE "THREE ROUND BODIES" AND THEIR MEASUREMENT

96. Solids with Curved Surfaces. In articles 15 and 89 geometrical solids and the plane and curved surfaces of solids have been defined, with examples cited, and in article 89 solids are classified as those having plane surfaces, or faces, and those having curved surfaces. In this chapter will be discussed certain solids having curved surfaces.

A *curved surface*, it will be remembered, is a surface of which no portion is plane. As curved lines may have any curvature or shape, so also may a curved surface. A curved surface may be of indefinite extent, or it may be a *closed* surface; and a closed surface may have fixed, definite edges or boundaries or it may not. Also it may have a definite describable shape and size, or it may not. Thus the surface of a ball, an egg, a barrel, a section of a round straight log or tube, are solids having closed curved surfaces of definite shape and size.

Such a surface will enclose a definite portion of space, and so the solid will have a definite *volume*. If the curved surface can be cut open and spread or rolled out flat, as a plane, without stretching or tearing, or imagined as so flattened out, then it will have definite boundaries, or edges, straight or curved. It will thus be equivalent to a definite plane surface and so have a definite *area*. This equivalent plane area is called the *surface area* or simply the *surface* of the solid bounded by the curved surface.

If the curved surface cannot be flattened out on a plane without stretching, tearing or distortion (as an orange peel or the rubber shell of a hollow rubber ball) then it can be thought of as divided into an extremely great number of extremely small sections, each of which can be spread on a plane without distortion. The sum of the areas of these minute sections will then be the area of the original curved surface.

There are of course curved surfaces of all shapes and all manner of curvature, and the number of such forms is indefinitely great. There are certain curved surfaces, however, which are of very simple forms. Thus an egg, which is slightly larger at one end than the other is a very common surface called an *ovaloid*. An egg which is of the same shape and size at both ends, curving off uniformly from the middle toward both ends, is called a *prolate ellipsoid*. An orange which is smoothly round except for a slight flattening at the stem and bud is of the form of an *oblate ellipsoid*, or as it is sometimes called, a *spheroid*. The familiar figure of a smooth, well-formed pear, or an electric lamp bulb, is called an *apioid*. The most familiar of all curved surfaces is, of course, a smooth round ball or globe, called the *sphere*.

97. The "Three Round Bodies." The closed figures studied in elementary plane geometry are polygons and the circle. Of the polygons the most important are triangles and quadrilaterals, and of these the most important are the right *triangle* and the *rectangle*, and a right triangle is half a rectangle. A rectangle, right triangle, and circle are shown in Fig. 183, these being of course already familiar.

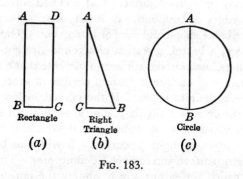

Fig. 183.

If the rectangle is rotated about the side AB as an axis the sides AD and BC as radii will trace out circles with centers at A and B, and the side CD will sweep out a *curved surface*, which is called a *cylindrical* surface. This curved surface is closed at the two ends by the circles formed by AD and BC, and the complete closed surface is called a *right circular cylinder*, or simply a *cylinder*.

If the right triangle (b) is rotated about a leg, say AC, as axis, the leg BC will trace out a circle, and the hypotenuse AB will sweep out a *curved surface*, which is called a *conical surface*. This curved surface is closed at a point (A) at one end and the other (open) end is closed by the circle formed by BC. The complete closed surface is called a *right circular cone*, or simply a *cone*.

If the circle (c) is rotated about any diameter, as AB, as an axis the boundary line of the circle sweeps out a uniformly curved, round, closed surface, which is called the *sphere*.

The surfaces or figures thus formed: *cylinder*, *cone* and *sphere*, are very important and familiar geometrical *solids* which are much used in the arts and sciences and in industry. They have many interesting and useful properties and have attracted the attention of geometers from the earliest times. The ancients called them "the three round bodies" and they still go by this collective name, though there are of course other round bodies.

In the remainder of this chapter we develop the rules for calculating the surface areas and volumes of the *three round bodies*.

98. The Cylinder. The figure formed by the rotation of a rectangle about one side, held stationary as an axis, as described in the preceding article is a *cylinder*. The two circular ends of the cylinder, turned out by the two ends of the rectangle as radii are called the *bases* of the cylinder. The radius of each of these circles is thus equal to the smaller dimension of the rectangle, and is called the *radius* of the cylinder. The side of the rectangle about which it was rotated is now a straight line through the center of the cylinder parallel to the curved surface and is called the *axis* of the cylinder. A cylinder is represented in Fig. 184. In this figure the line CD is the (extended) axis, and the length AB along the axis between the bases is called the *altitude* of the cylinder. AP is the radius and the curved surface S is called the *lateral surface* of the cylinder.

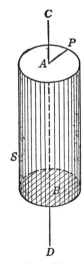

Fig. 184.

Examples of cylinders are: an ordinary tin can, a section cut out of a straight round rod, a length of round pipe, and a round unsharpened pencil. The length of such a cylinder is the altitude, the circular ends the bases; the diameter of the ends is the *diameter* of the cylinder, and their circumference the *circumference* of the cylinder.

If the curved lateral surface is rolled out flat, as the metal sheet of a cylindrical tin can or the paper label surrounding such a can, it will form a rectangle, one of whose dimensions is the altitude of the cylinder and the other the circumference of the cylinder. Since the area of this rectangle is the lateral surface of the cylinder and is equal to the product of its dimensions, then

    I. *The lateral surface of a cylinder is the product of its circumference and altitude.*

Now the circumference is, according to XXI, 84, 2π times the radius or π times the diameter. This rule becomes, therefore,

    II. *The lateral surface of a cylinder is 2π times the product of its radius and altitude; or π times the product of its diameter and altitude.*

Letting S stand for the lateral surface and using other symbols as before, these rules are written in symbols as

$$S = Ch = 2\pi Rh = \pi Dh.$$

The numerical value of π is usually taken to five figures, and is of course $\pi = 3.1416$.

The total surface area consists of the lateral surface and the two circular bases, and is, therefore, the sum of the lateral surface as just found plus twice the area of one of these circles. Using the formula for the area of a circle (XXIII, 84) we have for the total surface area

$$A = 2\pi Rh + 2\pi R^2 = 2\pi R(h + R),$$

or

$$A = \pi Dh + 2(\tfrac{1}{4}\pi D^2) = \pi D(h + \tfrac{1}{2}D).$$

    III. *The total surface area of a cylinder equals 2π times the radius times the sum of altitude and radius; or π times the diameter times the sum of altitude and half the diameter.*

If the altitude equals the diameter rule II above gives $S = \pi D \times D = \pi D^2$. But the area of the base is $B = \tfrac{1}{4}\pi D^2$. Hence $S = 4B$ and we have the remarkable result that

    II. (1) *If the altitude of a cylinder equals its diameter the lateral surface is four times the base area.*

Adding the two bases to the lateral surface we have also in case the altitude equals the diameter,

    III. (1) *If the altitude of a cylinder equals its diameter the total surface area is six times the base area.*

In order to calculate the volume of a cylinder we turn to the properties of prisms, article 92.

If in Fig. 176, the number of faces of the prism, and the number of sides of the base, be increased without at the same time increasing the over-all size of the base, that is, if the polygonal bases be thought of as inscribed in circles, the base polygons approach circles as limits (VII, 80), and the prism resembles more and more closely a cylinder. When the number of sides is extremely great and the width of each side or face extremely small, the prism is ultimately a cylinder. But the volume of the prism is the product of its altitude and base area. Therefore,

IV. *The volume of a cylinder is the product of its altitude and base area.*

But the base area of a cylinder is the area of a circle whose radius or diameter is that of the cylinder. Therefore,

V. *The volume of a cylinder is π times the product of its altitude and the square of the radius; or one-fourth π times the product of its altitude and the square of the diameter.*

In symbols these rules are written as the formulas,

$$V = \pi h R^2 = \tfrac{1}{4}\pi h D^2,$$

and as seen before $\tfrac{1}{4}\pi = 0.7854$, to four places.

As an illustration of the use of these rules, let us calculate the lateral surface, total surface area, and volume of a cylindrical tin can 3 inches in diameter and 10 inches high. Here the diameter is $D = 3$ and the altitude is $h = 10$. The lateral surface is, therefore,

$$S = 3.1416 \times 3 \times 10 = 94.25 \text{ sq. in.};$$

the total surface area is

$$A = 3.1416 \times 3 \times (10 + \tfrac{3}{2})$$
$$= 3.1416 \times 3 \times 11\tfrac{1}{2} = 108.39 \text{ sq. in.};$$

and the volume is

$$V = 0.7854 \times 10 \times 3^2$$
$$= 0.7854 \times 10 \times 9 = 70.69 \text{ cu. in.}$$

By working the rules backward, that is, by taking the formulas as ordinary algebraic equations and solving for D or R, the diameter or radius can be found when the altitude and the surface or volume are known; and similarly the altitude can be found when the diameter or radius and the surface or volume are known.

99. The Cone. We have seen (article 97) that if one leg of a right triangle is held stationary while the triangle is rotated about that leg as an axis, the hypotenuse sweeps out a curved surface which is called a right circular cone, or simply a *cone*. The other leg turns out a circle which is called the *base* of the cone; the radius of this circle is the rotating leg and is called the *radius* of the base of the cone. The curved conical surface is called the *lateral surface* of the cone. Fig. 185 represents a cone formed in this manner with the right △ABC. The line DE is the direction of the stationary leg AC and is called the *axis* of the cone. The point A is the *vertex* or *apex* of the cone, the circle with center at C is the base, and CB is the base radius. The length AC of the axis between the apex and the center of the base is the *altitude* and the length of the hypotenuse AB is the *slant height* of the cone.

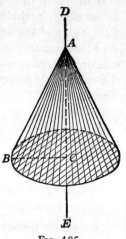

FIG. 185.

Examples of cones are: the round pointed end of a sharpened pencil, round pointed paper drinking cups, a round tapering pointed lamp shade, etc.

In order to find the surface area and volume of a cone we make use of the properties of pyramids (article 92). If in Fig. 178, the number of faces of the pyramid, and the number of sides of the polygonal base, be increased without at the same time increasing the over-all size of the pyramid, the pyramid resembles more and more closely a cone. If the base polygon be thought of as inscribed in a circle then as the number of sides is indefinitely increased the base polygon approaches the circle as a limit (VII, 80) and the pyramid ultimately becomes a a cone, the base perimeter of the pyramid becomes the base circumference of the cone, and its slant height, lateral surface and volume become the slant height, lateral surface and volume of the cone. By the rule for the lateral surface of the pyramid, therefore,

> VI. *The lateral surface of a cone is half the product of its base circumference and slant height.*

But the base circumference, according to the circle formula, is 2π times the radius. Using this relation the rule for the lateral surface of the cone becomes:

VII. *The lateral surface of a cone is π times the product of its radius and slant height.*

Letting s represent slant height and, as before, R the radius, this rule gives the formula
$$S = \pi R s.$$

Now the slant height AB in Fig. 185 is the hypotenuse of the right $\triangle ABC$, one leg of which is the altitude $AC = h$ and the other the base radius $BC = R$.

By the *Pythagorean Theorem*, therefore,

VIII. *The slant height of a right cone is the square root of the sum of the squares of its altitude and base radius,*

and in symbols,
$$s = \sqrt{h^2 + R^2}.$$

Using this expression for s the formula given above for the lateral surface S becomes
$$S = \pi R \sqrt{h^2 + R^2},$$
which expresses the lateral (curved) surface in terms of the dimensions (altitude and base radius) of the cone.

The total surface area of the cone is the sum of the lateral surface and the area of the circular base. Combining the rules for the lateral surface and for the area of a circle, therefore, the total surface area is
$$A = \pi R^2 + \pi R s = \pi R(R + s).$$

IX. *The total surface area of a cone is π times the base radius times the sum of the radius and the slant height.*

Using in this formula the square root expression found above for s it becomes,
$$A = \pi R(R + \sqrt{R^2 + h^2}).$$

In case the base diameter is known, half the diameter is the radius to be used in the rules and formulas for the cone.

Consider next the volume: since the volume is the ultimate volume of the pyramid, as explained above, then we have,

X. *The volume of a cone is one-third the product of its altitude and base area*, and since the base is a circle of radius R, then the volume is $V = \frac{1}{3}h \times \pi R^2 = \frac{1}{3}\pi h R^2$, or in terms of the diameter, $V = \frac{1}{3}h \times \frac{1}{4}\pi D^2 = \frac{1}{12}\pi h D^2$.

XI. *The volume of a cone equals one-third π times the altitude times the square of the radius; or one-twelfth π times the altitude times the square of the diameter.*

The numerical multipliers in these formulas are $\frac{1}{3}\pi = 1.0472$ and $\frac{1}{12}\pi = 0.2618$.

As an example of the use of the various formulas and rules for the cone let us find the slant height, lateral and total surface, and the volume of a cone 10 inches high and 6 inches across the base. Here the base diameter is 6 and therefore the radius is $R = 3$, and the altitude is $h = 10$ inches. By the formulas, therefore, the slant height is

$$s = \sqrt{3^2 + 10^2} = \sqrt{109} = 10.44 \text{ in.};$$

the lateral surface is

$$S = 3.1416 \times 3 \times 10.44 = 98.39 \text{ sq. in.};$$

the total surface area is

$$A = 3.1416 \times 3 \times (3 + 10.44)$$
$$= 3.1416 \times 3 \times 13.44 = 126.67 \text{ sq. in.};$$

and the volume is

$$V = 1.0472 \times 10 \times 3^2$$
$$= 1.0472 \times 10 \times 9 = 94.25 \text{ cu. in.}$$

100. The Sphere. A *sphere* is a globe, a perfectly round ball; a very familiar object and in form the simplest of the solids, just as the circle is the simplest of the plane figures. And as the circle is defined as a curved line every point of which is equally distant from a point inside it and in the same plane, so also may the sphere be defined as follows:

A *sphere* is a curved surface of which every point is equally distant from a point inside called the *center*.

Thus the center of a sphere is the "middle point" of its volume or of the space enclosed by its surface. Any line from the center to the surface, or the length of such a line, is a *radius* of the sphere. Any line through the center and having its two ends in the surface, or the length of such a line, is a *diameter* of the sphere. The diameter is therefore twice the radius, or the radius is half the diameter, as in the case of the circle.

As we have seen (article 97) the sphere may be generated by rotating a circle about a diameter as an axis. The circle therefore has the same diameter and radius as the sphere. It is called a *great circle* of the sphere and the circumference of the great circle is also the *circumference* of the sphere. Any circle having the same diameter and circumference is also called a great circle. If a sphere is cut straight through the center, it is divided into two equal parts and each part is called a *hemisphere*. If a great circle is drawn around the sphere on its surface (as the equator of the earth) the surface of the sphere is also said to be divided into two hemispheres.

If a sphere rotates about a diameter as an *axis*, the two ends of that diameter are called the *poles* of the rotating sphere and the circumference of a great circle whose plane is perpendicular to the axis at the center, that is, a line around the surface of the sphere midway between the poles, is called the *equator*. Any great circle passing through the poles is called a *meridian circle*.

Some of these definitions and dimensions are the same as those used in the discussion of latitude and longitude in article 58. As used in that article, the names and meanings are derived from the sphere as here described; they are applied to the earth because it is very nearly an exact sphere. Strictly, however, it is not quite exactly spherical, and is called a *spheroid*.

XII. *The surface area of a sphere equals the product of its diameter and circumference.*

For purposes of numerical computation this formula can be expressed in either of two more convenient forms. Thus since $C = \pi D$, $A = D \times \pi D = \pi D^2$; and also since $D = 2R$, $A = 2R \times 2\pi R = 4\pi R^2$. Thus

$$A = \pi D^2 = 4\pi R^2.$$

XIII. *The surface area of a sphere equals π times the square of the diameter; or 4π times the square of the radius.*

These are the formulas generally used in calculation. For bringing out more clearly the relation of the sphere to its generating great circle it can be expressed otherwise. Thus since R is the radius of the circle,

πR^2 is its area. The area of the sphere $4(\pi R^2)$ is then four times the area of the circle. That is,

XIV. *The surface area of a sphere is four times the area of its great circle.*

XV. *The volume of a sphere equals one-third the product of its radius and surface area.*

Like the surface area formula it may be expressed in several forms. Thus according to XIV the surface area equals four great circles of the sphere. The volume is therefore equal to $\frac{1}{3}(radius) \times (4\ grt.\ circles) = \frac{4}{3}(rad.) \times (grt.\ cir.)$.

XVI. *The volume of a sphere equals four-thirds the product of its radius and great circle area.*

These two propositions express the volume in terms of the surface and its relation to the generating circle.

For purposes of numerical computation, however, it is more convenient to express the volume in terms of the diameter or radius. This is easily done. For, the surface area is, by XIII above, $A = \pi D^2$, and $\frac{1}{3}$ the radius is $\frac{1}{6}$ the diameter. The formula of XV can therefore be written

$$V = \tfrac{1}{3}R \times A = \tfrac{1}{6}D \times \pi D^2 = \tfrac{1}{6}\pi D^3.$$

Also the area of the great circle is πR^2 and hence XVI gives

$$V = \tfrac{4}{3}R \times (circle) = \tfrac{4}{3}R \times \pi R^2 = \tfrac{4}{3}\pi R^3.$$

Writing these together,

$$V = \tfrac{1}{6}\pi D^3 = \tfrac{4}{3}\pi R^3.$$

XVII. *The volume of a cone is one-third that of a cylinder of the same base and altitude.*

To compare the volumes of a sphere and its circumscribed cylinder, we have (V, 98) for the volume of a cylinder $V_c = \tfrac{1}{4}\pi a D^2$ and when the altitude $a = D$, the diameter, then $V_c = \tfrac{1}{4}\pi D^3$. Also the volume of the sphere of the same diameter is $V_s = \tfrac{1}{6}\pi D^3$. Hence their ratio is

$$\frac{V_s}{V_c} = \frac{\tfrac{1}{6}\pi D^3}{\tfrac{1}{4}\pi D^3} = \frac{\tfrac{1}{6}}{\tfrac{1}{4}} = \tfrac{4}{6} = \tfrac{2}{3}.$$

101. Exercises and Problems.

1. Find the lateral and total surface and the volume of a cylinder one foot in diameter and five feet long.

2. What are the corresponding values for a cone of the same base and altitude as the cylinder in Example 1?

3. Enough paint is available to cover 60 square feet of the surface of a round column 15 inches in diameter. After one end is painted, how far along the length will the paint extend?

4. What are the total surface and volume of a cylinder whose diameter and altitude are the same?

5. The tin can which requires the least sheet tin for a given volume has its diameter and altitude equal. How many such cans may be made from 57.42 sq. ft. (disregarding waste) if each can is to hold a quart?

6. To measure the unknown volume of a solid of irregular shape it is immersed in water in a cylindrical vessel which is partly filled, and the resulting rise in level of the water is observed. Suppose the vessel has a diameter of $1\frac{1}{2}$ feet and the rise in level is 8 inches. What is the volume of the irregular solid?

7. How many cubic yards of earth or stone must be removed in constructing a tunnel a quarter of a mile long whose cross-section is a semicircle of 18 feet radius?

8. A conical pile of sand is 10 feet high and 16 feet across the base. How many loads does it contain, if the wagon box is $1\frac{1}{2} \times 3 \times 6$ feet and loaded level? How many square yards of tarpaulin are needed to cover the pile against rain?

9. The height of a cone equals its base diameter. What is the ratio of base area to lateral area?

10. What does the volume V of a cone become, if the altitude alone is doubled? If the base area alone is doubled? If the base area and altitude are both doubled?

11. Find the circumference, the great circle and surface area, and the volume of a 12-inch sphere.

12. If the ball on top of St. Paul's Cathedral in London is 6 feet in diameter, what would it cost to gild it at 7 cents per square inch, disregarding the support?

13. What is the numerical value of the diameter of a sphere if its surface area has the same numerical value as the circumference?

14. The volume of a sphere is 179 cu. ft., 1152 cu. in. Find its diameter and surface.

15. A 6-inch iron sphere is placed in a cylindrical can partly filled with water. If the diameter of the vessel is 8 inches, how much does the water level rise?

16. A hollow round ball has a diameter of 10 inches and weighs 75 pounds. Find the thickness of the metal shell. (See Ex. 3, article 99.)

17. A 6-inch glass globe is packed in a square (cubical) box, into which it just fits snugly. How much space in the box is to be filled with packing?

Chapter 11
NON-EUCLIDEAN GEOMETRY

102. Non-Euclidean Geometry. Euclidean geometry, which is based on the world of everyday experience, considered at first only three space dimensions and in that space takes as one of the fundamentals the notion of parallel straight lines. Euclid gave the definition: "Parallel lines are straight lines which, being in the same plane and being produced indefinitely in both directions, do not meet one another in either direction." He then stated as one of the foundation truths and principles that *Only one parallel to a given straight line can be drawn through a point outside that line.* This statement is called the Euclidean *parallel postulate*. For two thousand years mathematicians tried to prove or disprove this statement after the manner of the other propositions of geometry, but failed to do either. The idea finally began to prevail that since the statement can neither be proved nor disproved, it might be possible to build up a system of geometry which does not contain and does not depend on the parallel postulate.

In the course of the controversy which continued for twenty centuries an Italian priest and Professor of Mathematics at the University of Pavia, Girolamo Saccheri (1667–1733), published in 1733 a book whose title in English is *Euclid Freed from Every Fault*, in which he attempted to prove finally the parallel postulate. He proceeds in an unusual manner. He first develops all of Euclid's geometry which can be proved without the parallel postulate and extends it to new results in such a system. He then attempts to show that all these new results must be false and that therefore the postulate must be true. This second part of his work is not conclusive, however, and his original object fails. Without realizing it he really had written, in his first part, the first work on non-euclidean geometry. Saccheri's work was soon forgotten and remained unnoticed until comparatively recent years.

About a hundred years later there appeared in 1830 and 1832 two remarkable works which denied the necessity of the truth of the parallel postulate of Euclid and substituted in its place the postulate which states that *more than one parallel* to a line can be drawn through an outside point. Both built up the same complete new geometry on this basis, the first complete *non-Euclidean geometry*. A most remarkable thing about these two publications is the fact that their authors had never heard of each other and neither knew anything of the other's work. The first paper (1830) was written by a young Russian, Nicholas Lobachevski (1793–1856), Professor of Mathematics at the University of Kasan. Its title is *On the Principles of Geometry* and it was later rewritten and extended as *Researches on the Theory of Parallels*. The second work (1832) was written by a young Hungarian army officer, John Bolyai (1802–1860), and bears the title *The Science Absolute of Space*. Lobachevski and Bolyai are therefore the joint founders of non-euclidean geometry.

In 1854 a young German mathematician, Bernhard Riemann (1826–1866), presented as his graduation dissertation at the University of Göttingen a paper on the *Hypotheses which Lie at the Foundations of Geometry*, in which the principles of another non-euclidean geometry were set forth. Riemann also denied the Euclidean parallel postulate, but, contrary to the postulate of Lobachevski and Bolyai, he assumed that *no parallel* can be drawn. He went further and denied that a straight line can be produced indefinitely without returning upon itself, and also extended the notion of geometrical dimensions so that more than three might be included. Riemann's geometry has been developed as a differential geometry and has been found to be of the highest importance in Einstein's theory of relativity in which time plays the part of a fourth dimension, added to the three of Euclidean space.

An interesting and instructive little book on non-Euclidean geometry is *Non-euclidean Geometry: Three Moons in Mathesis*, by Lilian R. Lieber. In it references are given to larger works.

103. The Parallel Postulates. Let us first consider the parallel postulate of Euclid. Take a line SS' (Fig. 186) and a point B not on SS'. Euclid's postulate asserts that only one line parallel to a given

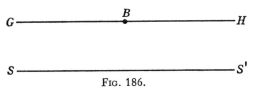

Fig. 186.

line can be drawn through a point outside that line; that is, through point B only the single line GH can be drawn that does not intersect line SS' at some point, however distant. (You will recall that in article 31 we said that parallel lines are those in the same plane that do not meet however far they might be extended.)

Next, consider a point F on line SS' (Fig. 187), and let us move that point to the right through points F_1, F_2, F_3, F_4 and draw lines through B to each point F on line SS'. As we move to the right, the lines consisting, in order, of BF, BF_1, BF_2, BF_3, and BF_4 move successively closer

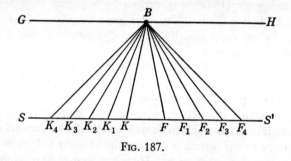

Fig. 187.

to that part of the line, GH, that is to the right of point B. By taking other points F still further to the right, the line drawn through B to such points will move closer to line GH as point F moves farther to the right. Similarly, we may take a point K on line SS' and move it to the left in the same manner in which we moved point F to the right. In this instance, as in the other, the lines drawn through point B and points K successively farther to the left will move closer to that part of line GH to the left of B.

In Euclidean geometry, it follows from the parallel postulates that however close lines BF and BK approach line GH or the respective sides of point B, these lines must *always* intersect line SS' at some point, however distant. They will fail to intersect line SS' only when they become identical with lines BH and BG, respectively. As we said earlier in this chapter, mathematicians spent nearly two thousand years trying to prove the truth of the Euclidean parallel postulate. Two distinct procedures for doing so were suggested: experimentally or mathematically. The first proves impossible, of course, for none of us has any knowledge of the distant reaches of space where such lines may or may not meet. On the other hand, after much effort to prove

NON-EUCLIDEAN GEOMETRY

that the Euclidean parallel postulate follows logically (that is, mathematically) from the other postulates, it was finally proved late in the last century that Euclid's parallel postulate is completely independent of the other postulates and cannot be proved by using them.

Consequently, mathematicians began to propose alternative postulates to Euclid's (which, for convenience, we shall speak of hereafter as Geometry E). We shall now consider these two alternatives.

A. That there are two lines through point B, not identical with BH and BG, which never intersect line SS' no matter how far extended. On Fig. 188 we shall call these lines BY and BZ. It

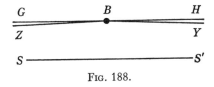

Fig. 188.

follows, moreover, that because line BY approaches but is not identical with BH, there is an angle of some unknown number of degrees between them (angle YBH). So long as this angle is greater than 0°, there is not only one line other than BH that does not meet line SS', but there may be an infinite number of such lines parallel to SS': that is, all those lines through $\angle HBY$ making an angle with BH greater than 0°, but of lesser number of degrees than that angle made with BH by BY. Thus, *there is more than one line parallel to a line through any point B outside that line.* This is the postulate substituted for Euclid's in the geometry of Bolyai and Lobachevski. We shall speak of this as Geometry L.

B. That there is a line through point B and some point F on line SS' that does eventually meet and become identical to line BH. Consequently, there is *no* line through point B which does not intersect line SS'. That is, *every* line through any point B not on line SS' will at some point meet SS'. Accordingly, every line in a plane must meet (Fig. 189) and *there are no parallel lines.* This is the geometry of Riemann. We shall speak of this as Geometry R.

The reader will object that the figures we have chosen to illustrate our point look wrong. For example, in Fig. 188, if line BY is extended sufficiently far it will surely meet line S'. This is so, but it is due to the inadequacy and necessary exaggeration of our drawing which arises from the size and, as we shall see, from the nature of the surface to which it is confined. Indeed, in the following consideration, of one of the more startling results of these non-Euclidean geometries, the reader must remember that the figures we use can only suggest and approximate the ideal figures. We must keep in mind, moreover, that we are accustomed to thinking of these matters in terms of Euclidean geometry, for which our ordinary drawings are quite satisfactory, but it does not follow that drawings of the same kind must necessarily comply with the other parallel axioms we shall be using.

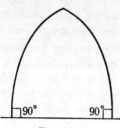

Fig. 189.

104. The Sum of the Angles in a Triangle. With this in mind and on the basis of these new postulates (Geometries L and R), we shall show that it is possible to take a set of postulates wholly or partially contradicting those of Euclid and build up a geometry as logically consistent as his. We shall now consider a brief *informal* proof of one of the striking results of substituting each of these new parallel postulates, in turn, for Euclid's. In this proof, the remaining postulates and definitions are as taken from the ordinary Euclidean geometry we have discussed throughout this book. The reader may wish to review the section on parallel lines in Euclidean geometry (article 31).

The following theorems are common to each of the three geometries:

Theorem A. If a triangle has two sides (AB and BC, in Figure 190) equal to two corresponding sides of another triangle, but the included angle of the first (ABC) is greater than the included angle of the second ($A'B'C'$), the third side of the first (AC) is greater than the third side of the second ($A'C'$). Conversely, if a triangle has two sides equal, respectively, to two sides of another triangle, but the third side of the first is greater than the third side of the second, the angle opposite the third side of the first is greater than the angle opposite the third side of the second.

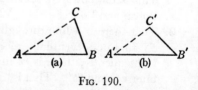

Fig. 190.

Theorem B. If two lines in a plane erected perpendicular to a third are unequal, the line joining their extremities makes unequal angles ($\angle ACD$ and $\angle CDB$) with them, the greater angle with the shorter perpendicular. (Figure 191(a).)

Theorem C. If the two angles at C and D are equal, the perpendiculars are equal, and if the angles are unequal, the perpendiculars are unequal, and the longer perpendicular makes the smaller angle. (Figure 191(b).)

We shall take the following theorems for Geometry L:

Theorem D. Two parallel lines move closer and closer to one another along their length and the distance between them becomes infinitely small.

Proof. Let AB and CD be parallel (Fig. 192) and from A and B, any points on AB, drop perpendiculars AC and BD to line CD. As-

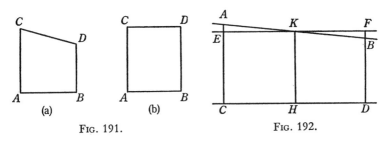

Fig. 191.   Fig. 192.

suming that B lies further to the right along the line AB, we shall prove that BD is less than AC. First, at the middle point of CD let us erect a perpendicular (HK) meeting AB at K. It follows that BKH is an acute angle and AKH is an obtuse angle. Thus, a perpendicular to HK at K must meet CA at some point between C and A and it will meet DB at some point F beyond B. But $DF = CE$; DB, therefore, is less than CA. By extending this to the right for successive portions of the parallel lines, we can prove that whatever the length of the line at some point similar to AC, the line bearing the same relation to it as BD is always of lesser length, and without limit.

Theorem E. Two non-parallel lines move farther from one another along their length and the distance between them becomes infinitely great.

Proof. Let us take two lines XY and XZ that intersect (Fig. 193). Take two points on XY such that XW is greater than XV and drop perpendiculars to XZ from points V and W. We shall prove that for a point W to the right of point V along XY, the distance of W from XZ is greater than the distance of V from XZ.

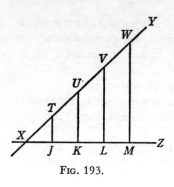

FIG. 193.

Consider, first, the possibility that WM and VL may be equal. If we were to draw a line from the middle point of LM perpendicular to XZ, such a line would also be perpendicular to XY. But we have said that these lines, XY and XZ intersect and it is impossible for the line to be perpendicular to the two lines if they intersect. Hence, WM cannot equal VL.

Consider, next, the possibility that WM may be less than VL. Take a line XT shorter than the length of either WM or XV and draw a perpendicular from point T to line XZ. Then, in accordance with the Pythagorean theorem, TJ is less than XT and both are less than WM. But we have said that VL is greater than WM. Thus, at some point along the line there must be a point U such that its perpendicular, UK, to XZ equals WM. But we have shown in the paragraph above that this is impossible.

Since WM can neither equal VL nor be less than VL, it follows that it must be greater. By treating successive portions of the line in this fashion it is possible to prove that for any line VL (that is, the distance between two non-parallel lines) there is another portion to the right where the perpendicular is of greater length, without limit.

Theorem F. When a transversal (see article 31) meets two lines making the sum of the interior angles on the same side equal to two right angles (= 180°), the lines do not meet and are not parallel.

Proof. Suppose AB is a transversal cutting XX' and YY' at A and B (Fig. 194) and that the sum of the angles XAB and ABY is equal to two right angles (= 180°). Hence, since angle ABY plus $Y'BA$ is also equal to two right angles, it follows that angle XAB is equal to angle $Y'BA$.

Now, bisect AB at point E, and draw ED perpendicular to XX' and EC perpendicular to YY'. It follows, then, that triangles EDA and ECB are congruent (see article 39, VI), and that angle $AED = CEB$. Therefore, DEC is a straight line perpendicular to XX' and YY'.

Consider, now, the possibility that X' and Y' eventually meet. If so, we can similarly prove that X and Y eventually meet. Next, consider the possibility that the lines $X'Y'$ are parallel to the right; if so, XY will be parallel to the left. However, according to Theorems D and E, above, these qualities cannot pertain to lines in Geometry L. Hence, they neither intersect nor are parallel. It follows that in

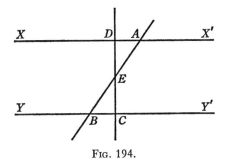

Fig. 194.

Geometry L, when a transversal meets two parallel lines, it makes the sum of the interior angles (XDC and YCD) less than the sum of two right angles. From the laws of symmetry they are equal to one another and are each, hence, less than 90°.

We shall take the following theorem for Geometry R:

Theorem G. All lines perpendicular to the same straight line meet at a given point outside the line and the distance between them becomes continually less until it reaches zero. This follows directly from the assumption that from any point outside a given line, no line can be drawn in the plane that does not meet that given line. (See Fig. 189.)

Keeping in mind the above theorems, as well as the others we have considered in this book, we shall now consider a particular property of triangles in each of these three geometries.

We shall now consider the following theorems:

Theorem H. The line joining the extremities of two equal perpendiculars may be equal to, greater than, or less than the line joining their opposite extremities (depending on which of the three geometries we consider).

Proof. Let AC and BD be the two such perpendiculars (Fig. 195). HK is a third perpendicular erected at the middle point of AB. Then,

HA and KC are perpendicular to HK, and KC is equal to, greater than, or less than HA depending on whether the angle at C is equal to, less than, or greater than the angle at A. This follows directly from Theorems B and C. Hence, CD, the double of KC may be equal to, greater than, or less than AB.

Fig. 195.

Corollary. If A, B, and D are right angles, the sides adjacent to angle C are equal to, greater than, or less than the sides opposite them, depending on whether angle C is a right angle, an acute angle, or an obtuse angle.

If possible, imagine in each of the figures below (Fig. 196 (*a*),(*b*),(*c*)) that the angles at A, B, A', B', A'', and B'' are all right angles. If so, then the angles at C and D are right angles (Geometry E, Fig. 196 (*a*)); C' and D' are acute angles (Geometry L, Fig. 196 (*b*)); and C'' and D'' are obtuse angles (Geometry R, Fig. 196 (*c*)).

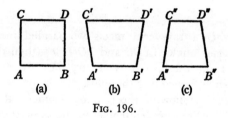

Fig. 196.

From the foregoing theorems, it follows that CD must equal AB in the case of the right angle, and that, in Geometry E, line CD remains the same length regardless of how distant it is from AB. You will recall, however, that in our non-Euclidean geometries, we said, in the case of Geometry R, that *all* lines must meet at a point; consequently, they continually move closer to one another until they meet. In this case, line CD is always less than AB and, according to common Theorems B and C, and Theorem G above, angle ACD is always obtuse (greater than 90°). In reference to Geometry L, however, we said that all non-parallel lines move farther from one another along their length. Here, specifically, line CD is always greater than AB, and, accordingly, the angle at C is acute (less than 90°), according to Theorems B, C, and F.

Theorem I. The sum of the angles of a triangle is equal to, greater than, or less than two right angles in the case of Geometry E, Geometry R, and Geometry L, respectively.

Proof. Take a right triangle *ABD* (Fig. 197) with right angle at *B*, draw *AC* perpendicular to *AB* and equal to *BD*. In triangles *ADC* and *DAB*, *AC* = *BD* and *AD* is common to each, but *DC* may be equal to, greater than, or less than *AB* in each geometry, respectively (Theorem G). Therefore, *DAC* is equal to, greater than, or less than *ADB* in each geometry, respectively. Add *BAD* to each angle, *ADB* plus *BAD* is equal to, less than, or greater than the right angle at *BAC*.

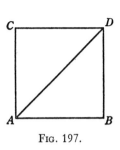

Fig. 197.

Accordingly, the sum of the angles in a triangle is

Equal to 180° in Euclidean geometry,
Greater than 180° in the geometry of Riemann, and
Less than 180° in the geometry of Lobachevski and Bolyai.

No less remarkable, we can even present an intuitively satisfying picture of each of these triangles and, for the geometry of Riemann, graphic evidence of two equally striking theorems that we have not attempted to prove here:

Of two triangles, the one with the greater angle sum has the larger area.

Two straight lines may enclose an area.

105. Visual Representation of Non-Euclidean Geometry. The triangle of less than 180° may be shown on a pseudosphere (Fig. 198).

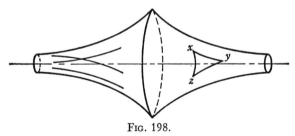

Fig. 198.

On such a surface—it has, incidentally, other specific mathematical properties—a curved line such as we have pictured is the shortest distance between two points. Because the postulates of Lobachevski and Bolyai apply to such a surface, it follows that the theorems derived from them must also do so, and we have pictured a triangle XYZ in which the sum of the angles is less than 180°.

Our illustration for the non-Euclidean geometry of Riemann is better known to us. It is the sphere.

Any lines on the surface of a sphere which are also on a great circle —that is, on a plane that cuts the center of the sphere—will be correctly described by Riemann's geometry. The triangle formed on line EF by C and D (Fig. 199) has two right triangles at C and D ($= 180°$) plus a third angle at A; hence, the sum of the angles in triangle ACD is equal to 180° plus the number of degrees in angle A. The reader will also see that points A and B are connected by *two* lines (through C and D, respectively) and that these two lines enclose an area.

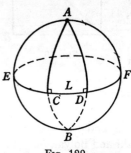

Fig. 199.

The reader may object that these lines on the sphere and pseudosphere are not "straight." But he will recall that we conventionally consider a straight line to be the shortest distance between two points. Our common usage of the term applies to plane surfaces, but there is nothing in the definition restricting it to such surfaces. On other surfaces, the shortest distance between two points is quite different from what we should expect. On a sphere, for example, it is along any line on a great circle, and the reader may test this for himself with a globe and a string held taut between various points.

No less important, these non-Euclidean geometries are not lacking in practical significance. Within relatively limited areas on the surface of the earth, we are safe in using our ordinary Euclidean geometry for our measurements and calculations. For larger areas, the geometry of Riemann has specific applications—boats and ships, for example, always choose a great circle along which to sail between any two points —and it has played a major role in the development of Einstein's Theory of Relativity. Equally important, our knowledge of the universe in which we live is still meager, and it is logically possible that our ultimate geometry may be non-Euclidean and best described in such terms. The writer hopes that he has made this seem more reasonable than it might have appeared at first glance.

REVIEW PROBLEMS

1. a. A 30°-60°-90° triangle has a hypotenuse of 10. Find the length of each leg.
   b. A square has a side of 3. Find the length of its diagonal in simplest form.
2. a. A 3"-4"-5" right triangle is rotated about the 3" leg to form a cone. Find the volume of the cone.
   b. Find the volume generated if the triangle is rotated about the 4" leg. Answers may be left in terms of π.
3. Given: $AB \parallel CD$.
   a. Name all pairs of alternate interior angles.
   b. Name all pairs of corresponding angles.
   c. Name all pairs of vertical angles.

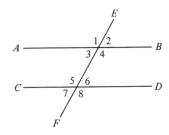

4. The angles of a triangle are in the ratio 2:3:4. Find the measure of the largest angle.
5. Find the measure of the angle between the hands of a clock when the time is 6:20.
6. How many 5" by 4" rectangular tiles are needed to cover a 3' by 5' floor?
7. A right circular cone is inscribed in a right circular cylinder of radius 4 and height 6. Find the volume inside of the cylinder but outside of the cone. Answer may be left in terms of π.
8. A circular wheel of radius 3 cm. rolls along the ground. How far has it traveled after 4 full rotations?
9. Find the length of the shadow cast by a 6-foot man at the instant when a 30-foot flagpole casts a shadow of 20 feet.
10. Can a circular mirror of radius $4\frac{1}{2}$ feet fit through a 4-foot by 8-foot doorway? Justify your answer.

11. A regular hexagon has side 10 mm.
    a. Find the length of the apothem of the hexagon. (Leave your answer in radical form.)
    b. Find the area of the hexagon. (Leave your answer in radical form.)
    c. Find the area of the circumscribed circle. (Leave your answer in terms of π.)
    d. Find the area of the inscribed circle. (Leave your answer in terms of π.)
12. The bases of an isosceles trapezoid are 9 and 15 and the length of a leg is 5. Find the area of the trapezoid.
13. Write TRUE if the statement is always true. In all other cases, write FALSE.
    a. All equilateral triangles are similar.
    b. All rectangles are similar.
    c. Two isosceles triangles are congruent if their legs are equal.
    d. The altitude to the base of an isosceles triangle bisects the vertex angle.
    e. The diagonals of a rhombus are equal.
    f. The diagonals of a rectangle are equal.
    g. If the side of a square is doubled, then its area is doubled.
14. A sphere has a radius of 10 inches. Find the area of the circle cut off when a plane intersects the sphere 8 inches from the center.

15. Given: right triangle PQR, median QM, $PQ = 12$, and $QR = 16$.

    Find the length of QM.

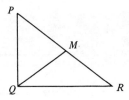

16. Given: triangle ABC, altitudes BG and CH, $AB = 16$, $CH = 6$, and $AC = 12$.

    Find the length of BG.

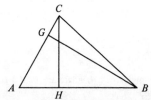

17. Find the surface area of a regular octahedron whose edge is 6 cm. Leave your answer in radical form.

18. A retangular solid has dimensions 3" by 4" by 6".
    a. Find the surface area of the solid.
    b. Find the volume of the solid.
    c. Find the length of a diagonal of the solid (AG). Leave your answer in radical form.

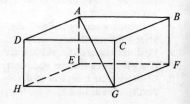

REVIEW PROBLEMS

19. Given: triangle *ABC*,
    line segment *B'C'*.

    Construct a triangle *A'B'C'*
    such that △ *A'B'C'* ∼ △ *ABC*.
    Use compasses and straight-
    edge only.

    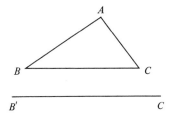

20. The longer leg of a right triangle is 1 inch less than twice the shorter leg. The hypotenuse is 9 inches more than the shorter leg. Find the lengths of all three sides of the triangle.
21. The lateral surface area of a cylinder of height 3 cm. is 24π square cm. Find the volume of the cylinder in terms of π.
22. A cube is inscribed in a sphere of volume 288π. Find the volume of the cube. (Leave your answer in radical form.)
23. Find, correct to the nearest tenth of a millimeter, the area of a flat washer whose inner radius is 3 mm and whose outer radius is 8 mm. Use $\pi = 3.14$.
24. A running track is in the shape of the accompanying diagram. The straightaways are 95 meters long and the radius of each semicircular end is 40 meters. Find the length of the entire track correct to the nearest meter if $\pi = 3.14$.

    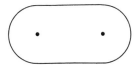

25. Given the figure with measurements as shown.

    a. Find the total surface area of the solid.

    b. Find the volume of the solid.

    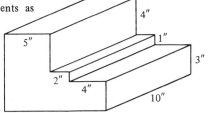

26. A slide projector uses a 1" by $1\frac{1}{2}"$ negative and projects it onto a screen 12 feet away from the negative while the light source is 6 inches away from the negative. Find the overall dimensions of the projected picture.
27. The circumference of a circle is 12π inches.
    a. Find the area of an equilateral triangle inscribed in the circle. (Leave your answer in radical form.)
    b. Find the perimeter of the triangle. (Leave your answer in radical form.)

28. Given: circle O, radii OA and OB, $\angle AOB = 120°$, $OA = 12$ mm.
    a. Find the length of $\widehat{AB}$ in terms of π.
    b. Find the area of sector AOB in terms of π.
    c. Find the area of triangle AOB correct to the nearest tenth of a square millimeter. Use $\sqrt{3} = 1.73$.
    d. Find the area of the shaded portion correct to the nearest square millimeter. Use $\pi = 3.14$.

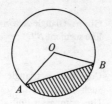

29. A "trefoil" (see figure at right) is made up of major arcs of circles drawn on an equilateral triangle with each vertex as center and half a side as radius. If a side of the triangle is 18 cm, find the total area enclosed by the trefoil. (Leave your answer in terms of π and radicals.)

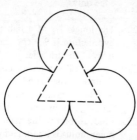

30. Given: rhombus $ABCD$, diagonals $AC = 10$, $BD = 24$.
    a. Find the area of rhombus $ABCD$.
    b. Find the length of a side of $ABCD$.
    c. Find the perimeter of rhombus $ABCD$.

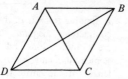

31. The volume of a cube is 64 cubic centimeters. Find the surface area of the cube.
32. a. Find the number of inches in the radius of a circle if the number of square inches in its area is the same as the number of inches in its circumference.
    b. Find the number of inches in the radius of a sphere if the number of cubic inches in its volume is the same as the number of square inches in its surface area.
33. The bases of a trapezoid are 11 and 19 and its area is 75.
    a. Find the length of an altitude of the trapezoid.
    b. Find the length of the median of the trapezoid.
34. Given: AB is tangent to the circle at E, secant BC intersects chord ED at point P, chord CD is drawn; $EF = 100°$, $EC = 150°$, $CD = 70°$. Find the measure of each of the following:
    a. FD
    b. $\angle 1$
    c. $\angle 2$
    d. $\angle 3$
    e. $\angle B$

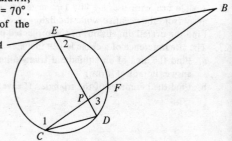

35. Given: PD is tangent to the circle at A, PE is tangent to the circle at B, $m \angle P = 20°$.

    Find $\widehat{AB}$ and $\widehat{ACB}$.

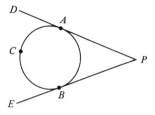

36. The supplement of an angle is 6° less than three times its complement. Find the number of degrees in the angle.

37. Given: triangle ABC, PQ ∥ BC, $AC = 28, AP = 8, PB = 6$.

    Find the length of AQ.

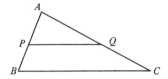

38. a. Find the sum of the exterior angles of a regular octagon.
    b. Find the measure of each exterior angle of a regular octagon.
39. The sum of the interior angles of a regular polygon is 1800°. Find the number of sides of the polygon.
40. Find the measure of each interior angle of a regular polygon of 10 sides.

41. Given: chords AC and BD intersect at point E inside the circle, $AE = 8$, $AC = 11, DE = 6$.

    Find the length of BE.

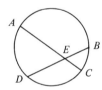

42. Triangle ABC is similar to triangle DEF. Corresponding sides $AB = 2$ and $DE = 3$. If the area of triangle ABC is 20, find the area of triangle DEF.

43. Given: In triangle ABC, $\angle ACB$ is a right angle, CD is an altitude.
    a. If $AD = 3$ and $DB = 12$, find the length of CD.
    b. If $AD = 4$ and $DB = 12$, find the length of AC.

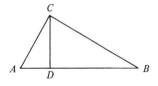

44. Given: point A on line l. Using compasses and straightedge, construct an angle of 45° on line l at point A.

45. A "golden rectangle" is a rectangle for which the ratio of its width to its length is the same as the ratio of its length to the sum of its width and its length. This rectangle's proportions are thought to be the most pleasing to the eye. Find, correct to the nearest integer, the length of a golden rectangle whose width is:

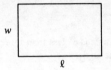

a. 3
b. 5
c. 8 (Use $\sqrt{5}$ = 2.2.)

46. The diagram shows a street map of Math City, U.S.A. Trig Drive is perpendicular to each of its cross streets, and the total length of Algebra Hill is 335 yards. Find the length of each of the three blocks on Algebra Hill.

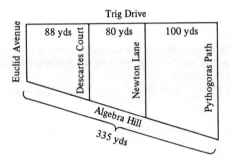

47. Given the following pairs of triangles with information as marked, which pairs of triangles are congruent? Justify your answer.

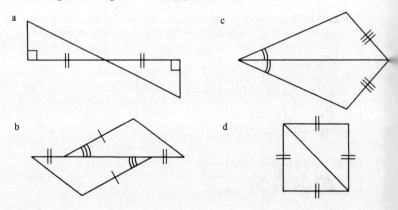

48. Given the following pairs of triangles with information as marked. Which pairs of triangles are similar? Justify your answer.

a

c

b

d

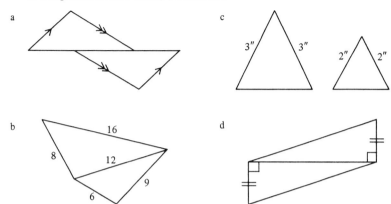

49. Given: trapezoid $ABCD$, $\angle A$ and $\angle D$ are right angles, $AB = 10$, $BC = 13$, $DC = 15$, $AD = 12$.

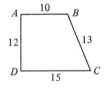

a. Find the volume of the solid generated when the trapezoid is rotated about side DC.
b. Find the volume of the solid generated when the trapezoid is rotated about side AB.
c. Find the volume of the solid generated when the trapezoid is rotated about side AD.

(Leave all answers in terms of π.)

50. A rectangular solid has dimensions 6" by 6" by 12". A cylinder of radius 2" is bored out of the solid by a drill bit perpendicular to the square face. Find the volume of the remaining material. (Leave your answer in terms of π.)

ANSWERS TO PROBLEMS

1. a. $5, 5\sqrt{3}$.
   b. $3\sqrt{2}$.
2. a. 16π.
   b. 12π.
3. a. $\angle 4, \angle 5; \angle 3, \angle 6$.
   b. $\angle 1, \angle 5; \angle 2, \angle 6$;
      $\angle 3, \angle 7; \angle 4, \angle 8$.
   c. $\angle 1, \angle 4; \angle 3, \angle 2$;
      $\angle 6, \angle 7; \angle 5, \angle 8$.
4. $80°$.
5. $70°$.
6. 108.
7. 64π.
8. 24π cm.
9. 4 ft.
10. No. $\sqrt{80} < 9$.
11. a. $5\sqrt{3}$ mm.
    b. $150\sqrt{3}$ sq. mm.
    c. 100π sq. mm.
    d. 75π sq. mm.
12. 48.
13. a. TRUE.
    b. FALSE.
    c. FALSE.
    d. TRUE.
    e. FALSE.
    f. TRUE.
    g. FALSE.
14. 36π sq. in.
15. 10.
16. 8.
17. $72\sqrt{3}$ cc.
18. a. 108 sq. in.
    b. 72 cu. in.
    c. $\sqrt{61}$ in.
20. 8, 15, 17.
21. 48π cc.
22. $192\sqrt{3}$.
23. 172.7 mm.
24. 441 m.
25. a. 500 sq. in.
    b. 600 cu. in.
26. 24 in. by 36 in.
27. a. $27\sqrt{3}$ sq. in.
    b. $18\sqrt{3}$ in.
28. a. 8π mm.
    b. 48π sq. mm.
    c. 62.3 sq. mm.
    d. 88 sq. mm.
29. $\dfrac{405\pi}{2} + 81\sqrt{3}$ sq. cm.
30. a. 120.
    b. 13
    c. 52.
31. 96 sq. cm.
32. a. 2 in.
    b. 3 in.
33. a. 5.
    b. 15.
34. a. $40°$.
    b. $20°$.
    c. $70°$.
    d. $95°$.
    e. $25°$.
35. $160°, 200°$.
36. $42°$.
37. 16.
38. a. $360°$.
    b. $45°$.
39. 12.
40. $144°$.
41. 4.
42. 45.
43. a. 6.
    b. 8.

45. a. 5.
    b. 8.
    c. 13.
46. 110 yds., 100 yds., 125 yds.
47. a. Yes; by angle = side = angle.
    b. Yes; by side = angle = side.
    c. No; side = side = angle is not a congruence pattern.
    d. Yes; by side = side = side.
48. a. Yes; angle = angle.
    b. Yes; all three sides are proportional.
    c. No; the included angles could be anything.
    d. Yes; if two triangles are congruent, then they are similar.
49. a. $1,680\pi$.
    b. $1,920\pi$.
    c. $1,900\pi$.
50. $(432 - 48\pi)$ cu. in.

ANSWERS TO EXERCISES

Article 28, Page 40
1. 84° 5′.
2. 16° 18′.
3. 84° 40′.
4. 73°.
5. Half of 90°; half of 45°.
6. As the sum of the angles is 180° the sum of their halves is 90°.
7. Yes.
8. 15°.
9. 20°.
10. 210°.
11. 24°.
12. 720°.
13. 6.

Article 35, Page 56
1. Zero degrees.
2. As the blade does not bend, each point on its edge moves through the same distance when the head is moved.
3. 145°.
4. (a) 30°; (b) 60°.
5. In order named: 140, 40, 140, 40, 140, 140, 40, 50, 50.
6. $b = c = 38°, d = 52°$.
9. (a) 21,600; (b) 1.153; (c) 69.2; .0192.
10. Very nearly 7930 miles.

Article 44, Page 84
1. Should be 180 degrees.
2. Yes; 45°.
3. No. With one right angle, each cannot be 60°.
4. Yes; twice as long.
5. 132°.
6. 120°.
7. See proposition III, 39.
8. Yes; 10 inches; about $9\frac{1}{2}$ inches.
12. 6.7 inches.
13. $13\frac{3}{4}$ inches.
14. *Suggestion:* Draw a figure like Fig. 44(a). Bisect $\angle A$ and extend the bisector to meet side BC at a point D; and similarly let the bisector of $\angle B$ meet AC at E. In triangles EAB and DBA, show by the construction and VII, 39, that $\angle EAB = \angle ABD$, and then by V, 39, that triangles EAB and DBA are congruent.
15. *Suggestion:* Sum of the adjacent interior and exterior angles is 180°; hence sum of their halves is 90°.
16. No; total would then be more than 180°.

Article 48, Page 96
1. See proposition XXIX, 46.
4. $BE = CE = 3, DE = 5$ inches.

6. 6 and 4 inches; yes; no.
7. *Suggestion:* Use XXXV, 46, and show sides equal; then use XI, 39.

Article 51, Page 103

No. of sides	Sum of int. angles	Each int. angle if regular
3	180°	60°
4	360	90
5	540	108
6	720	120
7	900	$128\frac{4}{7}° = 128° \ 34' \ 17\frac{1}{7}''$
8	1080	135
9	1260	140
10	1440	144
11	1620	$147\frac{3}{11}° = 147° \ 16' \ 21\frac{9}{11}''$
12	1800	150
15	2340	156

Article 59, Page 131

1. 90°; 120°; 75°; 105°.
2. 6000 deg. per sec.; 300 r.p.m.
3. $\frac{1}{4}$; $\frac{1}{2}$; 6; 360 deg. per min.
4. The second.
5. See proposition X, 54.
6. 180°.
7. 2.
8. 120°; see proposition XVII, 56.
9. 180°.
10. On the circle; smaller, 60°; larger, 120°.
11. 81°.
12. About 2200 miles.
13. $\angle CDO = COD = 25°$; $OCD = 130°$; $OCA = OAC = 50°$; $AOC = 80°$.
14. *Suggestion:* Use IV, 54.

Article 68, Page 155

1. 2; $1\frac{1}{2}$; 2.25; .75.
2. 2 : 5 or $\frac{2}{5}$ or .4.
3. $AD = 5\frac{17}{19}$, $BD = 8\frac{2}{19}$ inches.
4. 1, $1\frac{1}{2}$, 2 inches.
5. Draw a straight line from one end of a ruled line to the opposite end of the third line below it; through the points where this transversal crosses the two intermediate lines, draw lines parallel to the edges.
6. *Suggestion:* Through P draw a line $PD \parallel BC$ meeting AB at D. Then lay off $DE = DB$. Finish the construction and prove by IX, 62.
7. 12; 16.
8. Apply XXVII, 66, to each side by finding the fourth proportional to 3, 5, and the side.
9. 60 feet.
12. *Similarity* applies to shape alone; *congruence* involves both shape and size.
13. 8; 10 inches.
14. 1 : 9.
15. $3\frac{3}{4}$ inches.
16. $3\sqrt{5} = 6.71$ inches.
17. $19\frac{1}{3}$ inches.
18. About 38.4 miles.
19. 2.35 miles.
20. 2.83 mi.
21. 5.51 mi.
22. About 167 mi.

ANSWERS TO EXERCISES   257

Article 78, Page 180

1. 1215.
2. Smaller $\frac{3}{5}$ of larger.
3. Compare sides of △; 15 sq. in.
4. 5 in.; 70 sq. in.
5. No; yes.
6. 30 sq. in.
7. 30 sq. in.; 13 in.
8. 245 sq. in.; 15.65 in.
9. Draw diagonals of square and compare triangles formed with original triangle.
10. One sq. ft.
11. 1.79 acres.
12. $A = 6\sqrt{6} = 14.7$ sq. in.; $d = \dfrac{35}{2\sqrt{6}} = 7.15$ in.
13. 7.81 ft.
14. 42.42 yds.
15. 80 ft.

Article 87, Page 200

1.

No. sides	Central angle
3	120°
4	90
5	72
6	60
7	$51\frac{3}{7}° = 51° 25' 43'' -$
8	45
9	40
10	36
11	$32\frac{8}{11}° = 32° 43' 38'' +$
12	30
15	24

2. Larger 9 times smaller.
3. 482.8 sq. in.
4. 93.53 sq. in.
5. 597.4 sq. in.
6. 183 sq. in.
7. 11.76; 6.18; 4.16 in.
8. 3.16; 2.09 in.
9. 3.1416 ft.
10. 12.57 in.
11. 19 in.
12. 34.8 sq. in.
13. (1) 6.2832 in.; 3.1416 sq. in. (2) .7071; 1.414 in.
    (3) .9237; .7653 in. (4) 2; 2.8277 sq. in.
    (5) .3139 sq. in. (6) .8277 sq. in.
14. $\dfrac{\pi}{2}$.
15. $\dfrac{\pi}{4}$.
16. Four times as long.
17. 1.94; 6.11 in.
18. 720.
19. $\dfrac{11}{3\pi} = 1.165$ ft. $= 14$ in.
20. 805 sq. ft.
21. 21,204 lb. = 10.6 tons.
22. $\sqrt{2} = 1.414$ times as great.
23. 2696 deg. per sec.; 46.94 radians per sec.
24. About 1047 and 907 miles per hour.

Article 95, Page 221

1. 432 sq. in.; 576 cu. in.
2. 12.44 sq. yd.; 717.05 gal.
3. 6240 lb.; 720 sq. ft.
4. 2160.
5. 41.57 sq. ft.; 28 ft.; 388 cu. ft.
6. 96.
7. 8.32 ft.; .494 as great.
8. 6.928; 24; 13.856; 82.583; 34.641 sq. in.; .943; 8; 3.768; 61.304; 17.456 cu. in.
9. Areas as 1 : 2 : 5; volumes as 1 : 4 : 18.5.
10. Edge tetrahedron 2.04 times edge of cube.
11. 4.078.
12. 9.118 in.

Article 101, Page 232

1. 15.71; 17.28 sq. ft.; 3.93 cu. ft.
2. 7.85; 8.64 sq. ft.; 1.31 cu. ft.
3. About 15 ft.
4. Surface $\frac{2}{3}\pi \times$ (square of diameter). Volume $\frac{1}{4}\pi \times$ (cube of diameter).
5. 100.
6. 2036 cu. in. = 1.18 cu. ft.
7. 24,882 cu. yd.
8. Very nearly 25; 35.8 sq. yd.
9. $\sqrt{5}$.
10. $2V$; $2V$; $4V$.
11. 37.7 in.; 113.1, 452.4 sq. in.; 904.8 cu. in.
12. $1139.85.
13. One.
14. 7 ft.; 154 sq. ft.
15. $2\frac{1}{4}$ in.
16. 2.13 in.
17. 102.8 cu. in.

Answers to Review Exercises, Page 245

7. a. 59°.
   b. 28°.
11. 144 square inches.
12. 48.
13. Radius = 10 inches.
    Area of triangle = 48 square inches.
14. $81 - \frac{81\pi}{4}$ inches.
15. Arc $BC = 50°$.
    Angle $BCP = 25°$.
    Angle $ACB = 93°$.
16. a. 12.
    b. 36.
    c. 540.
17. a. $13\frac{1}{2}$ feet. b. 53 feet.
18. b. 7 inches. c. 33 inches. d. 94 inches.
19. a. 2π inches.
    b. $9\sqrt{3}$ square inches.
    c. 6π square inches.
    d. $(6\pi - 9\sqrt{3})$ square inches.
20. 15 square units.
22. $3\sqrt{3}$.
23. $75\sqrt{3}$.
28. $(100 - 25\pi)$ square inches.
33. 48π.
35. 40 inches.
40. a. $V = \frac{3e^2 h}{2}\sqrt{3}$.
    b. 840 pounds.
41. 457 cubic feet of cement.
42. a. 100.
    b. 65.
43. $32\pi\sqrt{3}$.
44. 93 pounds.
45. 58 cubic units.
47. 461 cubic feet.
48. $18\sqrt{2}$.
49. a. 72 square inches.
    b. $24\sqrt{3}$ cubic inches.
    c. 3π square inches.

INDEX

Altitudes, properties of, 68
Angles, 32
  drawing of, 51
  extension of the meaning of, 36
  measurement of, 38, 115
  plane figures formed by, 58
  properties of, 50
  relations, 35
  right angles, 34
Arcs, properties of, 107

Bisectors, angle, 68
  perpendicular, properties of, 68

Chords, properties of, 107
Circle, 104
  constructions, 120
  definitions pertaining to, 105
  finding diameter of, 152
  laying out a large, 129
  properties of, 104
  proportions, 142
  to find from arc, 128
Cone, 228
Cylinder, 225

Distances, inaccessible:
  measurement of, 148

Figures:
  plane, dimensions of, 161
  similar, 137
  similar geometrical, 138

Geometry:
  divisions of, 17
  fundamental definitions, 22
  general postulates of, 27
  illustrations and applications of, 76
  non-Euclidean, 234
Groove, to test a semi-circular, 128

Heights, inaccessible:
  measurements of, 150

Illustrations and applications of:
  the angle of elevation, 125
  geometry, 76
  latitude and longitude, 126
  proportions and similar figures, 146
  the protractor, 125

Latitude:
  illustrations and applications of, 126
  determination of, 127
Lines:
  in space, 203
  oblique, 40
  parallel, 46
    drawing of, 51
  perpendicular, 34
  straight, 31
Longitude:
  illustrations and applications of, 126

Measurement of:
  angles, 115
  area, 167
  inaccessible distances, 148
  a parallelogram, 167
  a rectangle, 163
  size of earth, 54
  a square, 166
  a triangle, 167
Medians, properties of, 68
Moon, eclipse, of 130

Non-Euclidean geometry, 234
  parallel postulate, 235
  sum of angles in a triangle, 238
  visual representation, 243

Oblique lines, 40

Pantograph, the, 147
Parallelograms, areas of, 163
Perpendicular lines, 40

INDEX

Pictures, enlarged, 146
Planes, in space, 203
Polygons:
    areas of, 174
    properties of, 99
    regular and circular, 182
Postulates, 27
Prisms, 213
Proportion, ratio and, 133
Pythagorean theorem, 173
Pyramids, 213

Quadrilaterals, 85
    properties of, 87

Ratio and proportion, 133
Rectangles, areas of, 163
Review exercises, 245
Right angles, 34
Right triangles, 70
    area of, 169

Segments, line, proportional, 135
Solids:
    rectangular, 208
    regular, 216
Sphere, the, 230
    with curved surfaces, 223
Squares, areas of, 166
Straight lines, 31
    plane figures formed by, 58

Tangents, properties of, 107
Trapezoids, areas of, 174
Triangles:
    areas of, 167
    construction of, 73
    descriptions and definitions of, 61
    forms of, 59
    sides, properties of, 68
    similar, properties of, 138
    with equal bases and altitudes, 168

Volume and solids, 208